Ion Implantation, Sputtering and their Applications

Ion Implantation, Sputtering and their Applications

by

P. D. TOWNSEND
*School of Mathematical and Physical Sciences
University of Sussex, England*

J. C. KELLY
*Associate Professor of Physics
University of New South Wales, Australia*

N. E. W. HARTLEY
*Nuclear Physics Division
A.E.R.E., Harwell, England*

1976

ACADEMIC PRESS
LONDON   NEW YORK   SAN FRANCISCO
A Subsidiary of Harcourt Brace Jovanovich, Publishers

ACADEMIC PRESS INC. (LONDON) LTD.
24/28 Oval Road,
London NW1

United States Edition published by
ACADEMIC PRESS INC.
111 Fifth Avenue
New York, New York 10003

Copyright © 1976 by
ACADEMIC PRESS INC. (LONDON) LTD.

All Rights Reserved

No part of this book may be reproduced in any form by photostat, microfilm, or any other means, without written permission from the publishers

Library of Congress Catalog Card Number: 75-196983
ISBN: 0-12-696950-7

Printed in Great Britain by
ROYSTAN PRINTERS LIMITED
Spencer Court, 7 Chalcot Road
London NW1

Preface

The interaction of ion beams with surfaces allows one to contour surfaces and change their properties in a way which is not otherwise possible. For both economic and scientific reasons the techniques of ion implantation and sputtering are ideally suited to the miniaturization and preparation of planar electronic devices in the form of integrated circuitry. In addition, ion implantation has a major role in device manufacture and exploratory research for fields as diverse as superconductivity or optical waveguides. Because the field of ion implantation might, and indeed should, appeal to a whole range of disciplines our aim in the early chapters of the book is to introduce the subject from a relatively simple level and then in the closing chapters demonstrate the vast range of potential applications.

Our own interest in ion implantation stems from postgraduate lecture courses given by us at the University of Sussex (where, in conjunction with Salford University, we offer an MSc course specifically in the area of atomic collisions in solids), at the University of New South Wales and our own research interests at these Universities and at Harwell.

We are indebted to many people for helpful discussions, for permission to reproduce figures and not least to the Science Research Council for financial support which enabled us to work together.

Our thanks are also due to Linda Lammiman for her excellent typing of the manuscript and to Michael Kelly and Tom Sully for their skill in preparing the figures.

March, 1975
Sussex.

P.D.T.

Contents

Preface			v
Chapter 1	INTRODUCTION		1
1.1	Ion implantation		1
1.2	The advantages of ion implantation		3
1.3	Conclusion		5
Chapter 2	ION RANGES IN SOLIDS		7
2.1	Processes of energy loss		7
2.2	Classical collisions		8
2.3	Interatomic forces		11
2.4	Scattering during collisions		15
2.5	Electronic energy loss		21
2.6	Charge exchange losses		24
2.7	Ion ranges in amorphous solids		25
2.8	Ion ranges in compounds		31
2.9	Concentration profiles		31
2.10	Implantation masks		34
2.11	Experimental measurement of ion ranges		35
2.12	Recent measurements of ion ranges		39
2.13	Energy deposition and defect formation		39
2.14	Conclusion		43
Chapter 3	ION CHANNELLING		45
3.1	Introduction		45
3.2	Simple channelling theory		49
3.3	Motion in a channel		54
3.4	The comparison of theory and experiment		58
3.5	Ion ranges in channels		62
3.6	Channelling yields		63
Chapter 4	RADIATION DAMAGE		65
4.1	Introduction		65
4.2	Vacancies and interstitials		66
4.3	Displacement energies		74
4.4	Alternative processes in defect production and separation		75
4.5	Defect concentrations		77
4.6	Extended defects		78
4.7	Fission fragments and fission tracks		79
4.8	Dislocations		79
4.9	Bubbles and voids		84
4.10	Summary		86

CONTENTS

Chapter 5	DIFFUSION AND THERMAL ANNEALING	88
5.1	Diffusion and annealing	88
5.2	Backscattering measurements of diffusion	90
5.3	Diffusion in thin films	93
5.4	Stress assisted diffusion	95
5.5	The stoichiometry of thin films	96
5.6	Glancing incidence Rutherford backscattering	98
5.7	Interstitial diffusion	105
5.8	Ion implantation gettering of damage	106
5.9	The diffusion of helium in copper	107
Chapter 6	SPUTTERING	111
6.1	Introduction	111
6.2	Sputtering theory	112
6.3	The Thompson model of sputtering	115
6.4	The Sigmund model of sputtering	117
6.5	The angular dependence of the sputtering yield	126
6.6	The velocity spectrum of ejected atoms	126
6.7	Surface topography	128
6.8	A comparison of sputtering and chemical dissolution	132
6.9	Secondary processes in surface topography	136
6.10	Sputtering of crystalline solids	142
6.11	Transmission sputtering	143
6.12	Sputter deposition	143
6.13	Conclusion	145
Chapter 7	ION BEAM EQUIPMENT	148
7.1	Ion accelerators	148
7.2	Ion sources	149
7.3	Radiofrequency ion sources	151
7.4	The duoplasmatron or Von Ardenne source	153
7.5	Surface ionisation sources	157
7.6	Scandinavian sources	160
7.7	Harwell ion source	161
7.8	Multiply charged ion sources	164
7.9	Other ion sources	165
7.10	Beam handling	166
7.11	The sector magnet mass analyser	167
7.12	The velocity filter	170
7.13	Beam scanning	171
7.14	Target manipulation	173
7.15	Beam transport	175
7.16	The measurement of dose	176
Chapter 8	ANALYSIS TECHNIQUES WITH ION BEAMS	181
8.1	Introduction	181
8.2	Rutherford backscattering	181
8.3	Channelling	198
8.4	Resonant elastic scattering	207
8.5	Nuclear reaction analysis	210

8.6	Ion induced X-rays	218
8.7	Analysis by electron irradiation and sputtering	225
8.8	Conclusions	240

Chapter 9 THE APPLICATIONS OF SPUTTERING — 247

9.1	Introduction	247
9.2	Machining and figuring	248
9.3	Polishing and the reduction of strain and scatter	249
9.4	Sample preparation for electron microscopy and surface analysis	251
9.5	The tuning of quartz crystal resonators	252
9.6	Piezoelectric transducers	253
9.7	High definition etch patterns	253
9.8	Surface acoustic waves	256
9.9	Sputtering used in integrated optics	256
9.10	Biological applications	258
9.11	Sputter deposition	258
9.12	Conclusion	259

Chapter 10 APPLICATIONS OF ION IMPLANTATION AND ION BEAM ANALYSIS — 262

10.1	Introduction	262
10.2	Semiconductors	262
10.3	Phosphors	263
10.4	Chemical effects	266
10.5	Electrochemistry	267
10.6	Corrosion and oxidation studies	269
10.7	Location of passivating implants	271
10.8	Potentiostatic measurements	272
10.9	Implantation of impurities for lattice site studies	275
10.10	Depth profile of fluoride in tooth enamel	278
10.11	Analysis of varnish layers of violins	279
10.12	Integrated optics	280
10.13	Friction and wear	287
10.14	Superconductivity	292
10.15	Ion implantation and magnetic bubbles	294
10.16	Simulation studies of damage in reactor materials	296

Appendix I RANGE – ENERGY DATA 304

Appendix II TABLE OF Z, A AND NATURAL ABUNDANCE OF THE ISOTOPES . 317

Appendix III ELECTRONICS FOR SURFACE ANALYSIS 326

Index 329

CHAPTER 1

Introduction

1.1 ION IMPLANTATION

Ion implantation is conceptually one of the simplest methods for building a surface with particular properties in selected regions. In essence the entire problem is to decide which chemical, electrical, mechanical or optical properties are required in an area of the surface, then to choose the chemical composition which will achieve these results. A judicious choice of injection energies and masks will then allow the implanted impurities to determine the depth and spatial shape of the modified layers. In normal systems the depth scale of the change is limited to the order of a micron.

In this book we will describe some of our present understanding of the processes involved in ion implantation, both in amorphous and crystalline solids. We will also mention some of the side effects of implantation, the major one being the formation of radiation damage. Whilst this effect is a nuisance in predicting the property changes induced in implanted regions it may frequently be avoided or removed by a suitable choice of the temperature of the solid during or after implantation. In many situations the disorder created by the passage of fast ions can be entirely beneficial and can produce the property changes required and it is unfortunate that the phrase "radiation damage" is used, which implies an unwanted effect. For example diffusion rates may be enhanced in a "damaged" solid, optical waveguides may form in the "damaged" region or the surface may become corrosion resistant.

We will also devote a section to topographical changes produced by ion beam sputtering of the surface. This is normally a trivial effect in implantation studies but there are many conceivable applications for the machining of small surface features by ion beams. One can of course use the same accelerator system for both sputter machining and ion implantation by changing the ion and accelerator energy.

The chapters on the theory of ion ranges, the depth profile of the implant in amorphous and channelled situations, the experimental measurements of the ranges, and the current trends in the hardware of accelerators, mass separators and target chambers, are common to all materials, so we will overlap with the material in the books by Carter and Colligon (1968), Mayer *et al.* (1970), Dearnaley *et al.* (1973). In these books the major emphasis is on semiconductor physics. We have chosen to exclude most of the work on semiconductors and concentrate on the newer fields of implantation effects in metals and insulators. The existing literature in these fields is more limited but the future of both is diverse and important. Corrosion, fuel cells, optical waveguides and mechanical wear are but a few of the areas where ion implantation will make an impact and which are discussed in this book. Many of these experiments are still in their infancy and the quoted results must not be treated as the ultimate changes which may be produced. Instead it is intended to show that ion implantation is a much more versatile technique and there is immense scope for imaginative applications.

The precedent for the applications of the technology are clearly established in the semiconductor industry and it is instructive to record the early history in that area. In the early 1960's it was known that electrical properties of semiconductors could be changed both by radiation damage and the injection of fast ions. The research laboratory conditions and the variations in results discouraged commercial interest at that stage. Some of the uncertainties disappeared when it was realised that the simple range predictions in amorphous materials were inappropriate if the ions were directed down the crystallographic channels which exist in solids. A better understanding of the theory together with improvements in accelerator design and reliability meant that semiconductor devices could now be made by ion implantation. The existing expertise in diffusion produced devices was well developed so that the marginal advantage of control of the impurity profile from implantation was ignored. At this stage of the story it was conceded that ion implantation could be used for special high cost devices. However, the situation changed when implantation was used to make large batches of devices because the percentage of successful implanted circuits was far greater than diffusion made ones. This lowered unit cost made ion implantation an acceptable technology and by this stage one no longer comments that the integrated circuits of, for example, pocket calculators are entirely the result of implantation. However, we are still at a very early stage of the story for implantation changes to be accepted in metal or insulator technology and the semiconductor precedent would suggest that existing techniques can often be improved. Competition between different technologies is an excellent incentive for progress and we hope that this book will encourage not only advocates of ion implantation but also those who are challenged by it.

1.2 THE ADVANTAGES OF ION IMPLANTATION

Before we enter into the details of the theory or experimentation we shall first take a more general look at the inherent advantages of introducing impurity atoms at high velocity. Injection of ions from an accelerator implies that the new material is not formed in thermodynamic equilibrium. This is extremely valuable because normal chemical solubility rules are bypassed and it is quite possible to achieve impurity levels which are inaccessible by conventional treatments. In addition we can control the temperature of the solid during implantation so we may implant into crystal phases which would not form had the impurities been added at an earlier melt stage in the crystal growth. The random nature of the scattering and energy loss as the ions are slowed down in the solid results in an atomically dispersed system of impurity atoms for quite high dopant levels. But again because we can control the temperature of the solid both during and after the implant we can allow aggregation or diffusion of the impurities to proceed if we so desire.

In essence ion implantation should offer a simple method for the study of diffusion dynamics but in practice the measured diffusion coefficients for classical and ion implanted systems differ both because the diffusion rate is enhanced by the presence of radiation produced defects and also ion implantation is insensitive to grain boundaries and dislocations in the solid. Classical diffusion measurements may be distorted by precipitation or preferential migration along these regions of imperfection and the quoted diffusion coefficients would then be inappropriate for use in implantation studies.

More sophisticated advantages of diffusion techniques combined with ion implantation might include enhanced diffusion in certain regions of a solid by producing a pattern of radiation damage with one ion and then diffusing a second type of ion. The radiation enhanced diffusion would then transport the second ion along the pattern defined by the radiation damage.

Chemical doping from an ion beam is a "clean" method and with accelerators and mass separators of high resolution it is quite feasible to implant a single isotope. In a research situation this high mass selectivity may be useful for example in making a substance labelled with a specific radioactive isotope. In insulators, and semiconductors, the powerful techniques of electron spin resonance and electron nuclear double resonance can reveal the nature of atomic environments by the resonance patterns produced by electronic and nuclear spin interactions. Doping with a single isotope can frequently simplify these resonance patterns and aid the interpretation.

The simplicity with which atoms can be added to a solid also implies that a second or third impurity can be added under controlled conditions without problems of material decomposition or the accidental incorporation of unwanted impurities. In the manufacture of luminescent phosphors or photoemissive materials, and in semiconductors, one is frequently plagued by trace impurities which entered the solid during its preparation. They may be in too low a concentration to accurately analyse and vary from batch to batch. Ion implantation offers an alternative approach to the preparation of these materials, both at the research stage to develop the correct impurity and at the manufacturing stage to maintain quality control.

For new systems one should not ignore the ideal advantages of ion implantation for preparing a rapid survey of a range of alloys since the metallurgical problems of grain size and solubility are minimized. Examples of such exploratory uses of ion implantation will be given in Chapter 10.

In semiconductor physics one is frequently concerned with defect or impurity sites in concentrations of less than parts per million and we only consider the subtle control of properties which result from changes in the Fermi level. There is no *a priori* reason why the implanted impurities should not constitute several per cent of the total number of atoms in the surface region and so produce large effects on the chemical and physical properties near the surface. The present limitation is probably set by the strain associated with the presence of impurity ions and the fact that the implanted region only extends about a micron below the surface. The surface layer can then be considered as a totally new material which coats the original solid. The value of this concept is well-known but not always easy to achieve. Obvious examples include the desire to have soft materials for accurate machining which are case hardened or non corrosive to protect them during their working life. The two needs may not be compatible if the material distorts during the hardening process. However, the formation of a tough surface structure by ion implantation can avoid this problem. Even if the surface and bulk materials differ in atomic spacing there is no reason why the "coating" cannot be smoothly joined to the solid by control of the implanted ion profile. The result is a surface layer which is an integral part of the solid without the discontinuities or adhesion problems of a chemically added coating.

The new surface will differ in chemical stability from the original so ion implantation can be used either as a corrosion inhibitor or as a chemical reaction catalyst. Our present knowledge of corrosion processes in metals allows us to predict surface structures which would inhibit corrosion, however this knowledge should not obscure the fact that alternative alloys may be produced by implantation and these could be superior to those inhibitors which are already used. The art of corrosion resistance is far less developed with insulators and there are many substances which have desirable optical or electro-optic properties but are ignored because they are soft or hygroscopic. An ion implanted surface layer could provide the necessary protection.

If one wanted to utilise further the versatility of ion implantation then one could correct optical aberrations by varying the refractive index of the coating in selected areas of the surface. One such example could be a parallel sided single element lens.

Chemical instability in regions disordered by ion implantation is already used in semiconductor manufacture to change the sensitivity of photoresist material. The effects can be quite spectacular, for example it is reported that ion implanted garnet etches up to 2000 times faster than normal material. This selective etching is valuable because it avoids the sideways undercutting that occurs in normal masking techniques. Enhanced sputtering also occurs in previously damaged material. These effects allow high precision machining on insulators which offers the possibility of completely new technologies for these materials. Sputtering applications will be discussed in Chapter 9.

1.3 CONCLUSION

In this chapter we have sketched some of the areas where ion implantation can be valuable to indicate the versatility of the subject. In subsequent chapters we shall discuss the theoretical and experimental assessment of the implantation profiles, the consequences of radiation damage and the changes brought about by heat treatment of implanted specimens.

Many accelerator facilities are in research areas so the machines are also being used to produce ion beams for sputtering and as analytical probes. We therefore discuss topographical changes and the very powerful technique of Rutherford backscattering.

In order to gain acceptance in commercial processes it is desirable that ion implantation or sputtering facilities are seen to be efficient and reliable. Nevertheless this image is often distorted because existing machines which are operated in research areas may be excessively complex in order to achieve flexibility for many experiments. This image of accelerators is unfortunate since commercial machines for specific applications are simple to operate, reliable and produce large stable beam currents, as is evidenced by their use in semiconductor production. A description of some current machines is included in Chapter 7.

The final chapter provides a range of examples in which ion beams are being used to develop new materials. We sincerely hope that this section will both stimulate new ideas and encourage industries to use some of the results in production systems.

REFERENCES

Agajanian, A. H. (1974). *Rad. Eff.* **23**, 73.
Carter, G. and Colligon, J. S. (1968). "Ion Bombardment of Solids", Heinemann.

Crowder, B. L. (Ed.) (1973). "Ion Implantation in Semiconductors and Other Materials", (Yorktown Heights, 1972) Plenum Press.
Crowder, B. L. (Ed.) (1975). "Ion Implantation in Semiconductors and Other Materials", (Osaka, 1974) Plenum Press.
Dearnaley, G. (1969). *Rep. Prog. Phys.* **32**, 405.
Dearnaley, G., Freeman, J. H., Nelson, R. S. and Stephen, J. (1973). "Ion implantation", North Holland.
Degen, P. L. (1973). *Phys. Stat. Sol.* **A16**, 9.
Eisen, F. H. and Chadderton, L. T. (Eds) (1971). "Ion Implantation" (Proc. 1st Int. Conf.—Thousand Oaks), Gordon and Breach.
Gibbons, J. F. (1972). *Proc. IEEE*, **60**, 1062.
Mayer, J. W., Eriksson, L. and Davies, J. A. (1970). "Ion Implantation in Semiconductors", Academic Press.
Mayer, J. W. and Marsh, O. J. (1969). *Applied Solid State Science*, **1**, 239.
Namba, S. (Ed.) (1971). "Ion Implantation in Semiconductors" (Proc. U.S.-Japan Seminar), Japan Soc. for promotion of science, Kyoto.
Palmer, D. W., Thompson, M. W. and Townsend, P. D. (Eds) (1970). "Atomic Collision Phenomena in Solids", North Holland.
Picraux, S. T., Eer Nisse, E. P. and Vook, F. L. (Eds) (1974). "Applications of Ion Beams to Metals", Plenum Press.
Ruge, I. and Graul, J. (Eds) (1971). "Ion Implantation in Semiconductors" (Proc. 2nd Int. Conf.—Garmisch-Partenkirchen), Springer-Verlag.
Schulz, M. (1974). *Applied Physics*, **4**, 91.
Thompson, M. W. (1969). "Defects and Radiation Damage in Metals", Cambridge University Press.

General References

Carter, G. and Grant, W. A. (1976). "Ion Implantation in Semiconductors", Arnold.
Corbett, J. W. and Ianiello, L. C., (Eds) (1972). "Radiation Induced Voids in Metals", U.S.A.E.C. Symp. series 26.
Friedel, J. (1964). "Dislocations", Addison-Wesley.
Henderson, B. (1972). "Defects in Crystalline Solids", Arnold.
Kittel, C. (1971). "Introduction to Solid State Physics", Fourth Edition, Wiley.
Nabarro, F. R. N. (1967). "Theory of Crystal Dislocations", Oxford University Press.
Namba, S. and Masusla, K. (1975). *Electronics and Electron Physics,* **37,** 264.
Proc. of Amsterdam Conf. on "Atomic Collisions in Solids" (1976). "Surface Science", to be published.
Proc. of Warwick Conf. on "Ion Implantation in Semiconductors and Other Materials" (1976), I.O.P.
"Theory of Imperfect Crystalline Solids (1971). I.A.E.A., Vienna.

Chapter 2

Ion Ranges in Solids

2.1 PROCESSES OF ENERGY LOSS

If we are to control the properties of ion implanted surfaces then we must understand the mechanisms of energy loss which control the depth distribution of the implant as well as the radiation damage which ensues along the track as energy is transferred to the solid. To do this we consider the scattering events which influence the total path length, the range straggling and the projected depth of the implant in the direction of the original ion beam. The theories contain many assumptions and empirical correction factors so for practical applications one might wish to proceed directly to the widely used range estimates which are those of Lindhard, Scharff and Schiøtt (1963), (LSS). Range-energy tables have been computed from their theory for a variety of ion-solid systems by Johnson and Gibbons (1970) and Smith (1971). A selection of typical data is included in the books by Carter and Colligon (1968), Mayer et al. (1970), Dearnaley et al. (1973), Wilson and Brewer (1973) and in Appendix I of this present volume.

Calculations of energy deposition with depth into the target material will also give some measure of the radiation damage produced during the implantation. These depth distributions will be discussed in Section 2.12. It should be noted that the stability of the displaced atoms will determine the final state of the radiation damage and this is discussed in greater detail in Chapter 4.

We shall first discuss the derivation of the LSS range theory. The major processes of energy loss are (i) direct collisions between the ion and a screened nucleus, (ii) excitation of electrons bound in the solid and (iii) charge exchange processes between the ion and the atoms of the solid. All three processes are energy dependent and so make different contributions to the energy loss along the path of the ion. For simplicity we will treat the three independently and write a differential energy loss equation as

$$\left(\frac{dE}{dx}\right)_{loss} = \left(\frac{dE}{dx}\right)_{nuclear} + \left(\frac{dE}{dx}\right)_{electronic} + \left(\frac{dE}{dx}\right)_{exchange}.$$

Correlations between the events and the subsequent motion of the displaced atoms will be ignored whilst considering the range of implanted ions. However we must consider movements of the displaced atoms when we discuss radiation damage, as the displaced atoms can in turn impart sufficient energy to cause many further displacements. Fortunately the problem can be treated classically as a scattering event between two ions. In part this is possible because the dominant energy loss, by elastic nuclear collisions, occurs for ions between 5 and 500 keV and also detailed electronic structure effects are lost in amorphous solids because the final range is determined by many individual events. This averaging effect allows us to use the LSS theory for non-crystalline materials. Adjustments to the theory are required when the ion moves nearly parallel to crystal directions. Many crystalline directions act as channels and the ion bounces along with a relatively weak interaction with the walls, however in this instance specific electronic orbitals are involved so electronic shell effects must be considered. The associated reduction in dE/dx frequently produces channelled ranges up to ten times those expected for ions injected in a random direction. Channelling plays a significant role in ion implantation studies and light ion analytical studies of Rutherford backscattering (see Chapter 8).

2.2 CLASSICAL COLLISIONS

In the energy range used for implantation, 5–500 keV, the dominant energy loss is by elastic interactions between the ion and a screened nucleus. The problem is simplified by considering two body events in a centre of mass coordinate system to predict the energy loss, T, the cross section for energy transfer, $d\sigma_{nuclear}$, and the angle of scattering. The trajectory of the particles for a two body collision event is shown in Fig. 2.1 for a laboratory frame and the equivalent collision in a centre of mass frame is presented in Fig. 2.2. In the laboratory frame the momentum must be $M_1 u_1 = (M_1 + M_2) V_{CM}$. This is equivalent to moving the particle M_2 in the centre of mass frame at a velocity

$$U_2 = -\frac{M_1 u_1}{(M_1 + M_2)}.$$

Since the collision is elastic both energy and momentum are conserved so

$$M_1 U_1^2 + M_2 U_2^2 = M_1 V_1^2 + M_2 V_2^2$$

and

$$M_1 U_1 - M_2 U_2 = M_1 V_1 - M_2 V_2$$

2.2 CLASSICAL COLLISIONS

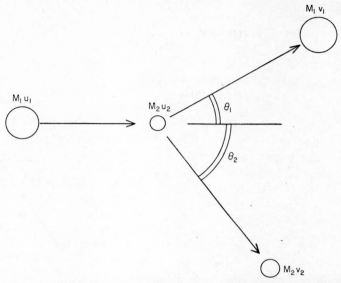

FIG. 2.1. A collision between two particles in a laboratory co-ordinate system.

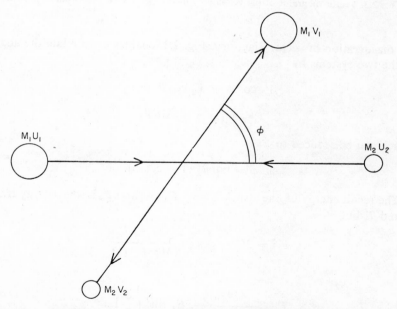

FIG. 2.2. A collision between two particles in a centre of mass frame co-ordinate system.

which implies

$$U_1 = V_1 = \frac{M_2 u_1}{M_1 + M_2}$$

and

$$U_2 = V_2 = \frac{M_1 u_1}{M_1 + M_2}.$$

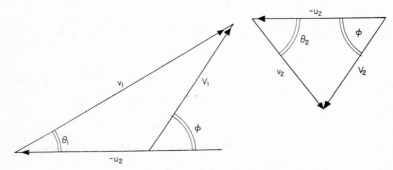

FIG. 2.3. Vector diagrams of the velocities which relate the two co-ordinate systems.

A consideration of vector diagrams, Fig. 2.3, enables us to relate the angles in the two systems by

$$V_2 \cos \phi + V_2 \cos \theta_2 = u_2$$

$$V_2 \sin \phi = V_2 \sin \theta_2$$

which can be reduced to

$$\sin \phi = \tan \theta_2 (1 - \cos \phi).$$

The recoil energy of the struck atom, $E_2 = \tfrac{1}{2} M_2 v_2^2$ is the energy transferred T, but

$$v_2^2 = 2(1 - \cos \phi) \frac{M_1^2 u_1^2}{(M_1 + M_2)^2}$$

so

$$T = \frac{4 M_1 M_2}{(M_1 + M_2)^2} E_1 \sin^2 \left(\frac{\phi}{2} \right).$$

This sets an upper limit for energy transfer in a head-on collision as

$$T_{max} = \frac{4M_1M_2}{(M_1+M_2)^2} E_1 .$$

E_1 and T may differ by several orders of magnitude if there is a disparity between M_1 and M_2 and only a small angle collision occurs, it is therefore sensible to ask if a classical approach is justified.

The moving particle has an associated wavelength $\lambda = h\sqrt{[1/(2ME)]}$ and is striking an object of radius a. The scattering will be classical if $\phi \gg \lambda/(2\pi a)$. If we compare the particle size, a, with the Bohr radius, $a_0 = \hbar^2/(m_0 e^2)$, where m_0, e are the electronic mass and charge, then

$$\frac{\lambda}{2\pi a_0} = \sqrt{\frac{e^2 m_0}{2a_0 ME}},$$

so classical mechanics will suffice if

$$E \gg \frac{e^2 m_0}{2a_0 M} \left(\frac{a_0}{a}\right)^2 .$$

In practice this low energy limit is a fraction of an electron-volt in energy so even for small angle collisions ion implantation problems can be treated classically.

2.3 INTERATOMIC FORCES

Having decided that we can consider classical elastic collision events we must now choose a form for the repulsive potential between the ions in order to assign a finite radius to the particles and hence predict a scattering angle. Ideally one would choose a simple analytical expression for $V(r)$ and historically the Coulomb, Bohr and Born–Mayer potentials were of this form. Improvements to these functions were generally made by modifying factors rather than new functions. However, the successful descriptions of $V(r)$ over a wide range of ion separations have been made by numerical means. Such functions are now of value since computations are feasible with large modern computers. Figure 2.4 compares some of the alternative repulsive potentials which have been used and examples of analytical descriptions are given in Table 2.1. It is evident that the agreement between the curves extends over a limited range of separations. There is no definitive potential which is appropriate for all pairs of ions and all energies so even the simplest expressions contain empirically adjusted parameters. Comparison with experiment is difficult because elasticity and compressibility experiments only allow one to sense the potential near the equilibrium atom spacing in the solid. The alternative is to probe the interaction in a dynamic event, i.e. by observing the scatter of an energetic ion beam. The principle is well established by the high energy alpha particle experiments of Rutherford (1911) to determine the size of the nuclei. Unfortunately ion implantation and radiation damage events involve the middle region of the potential curve which is the most difficult to characterise. One approach is to measure the scattering of ion beams from an inert gas and compare the results with an assumed potential. A recent review of this field was given by Kessel (1970).

An alternative approach is to use a computer simulation of a solid and compare ion range predictions with experiment. In principle one need only to assign initial positions to the ions, specify a repulsive potential between pairs of atoms and provide stability to the system by imposing a boundary condition which simulates the cohesive binding forces of the solid. When an energetic foreign atom "appears" in this system the lattice is perturbed and dynamic changes take place in the structure. However these can be followed by solving the classical equations of motion to find the new atomic positions after some small time increment. Iterative calculations of this type, coupled

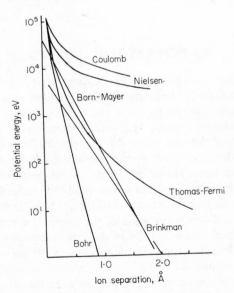

FIG. 2.4(a). A comparison of various proposed interatomic potentials between two copper atoms.

with computer printouts of the new positions, will display the development of radiation damage and the progress of the foreign ion. Suitable allowance for thermal energy will also show annealing of the radiation damage as the "lattice" relaxes to an equilibrium condition. Many events simulated in this way will produce a statistical view of the implantation process.

Such a computer program has the advantages that, once written, it allows one to explore variations in potentials or defect stability and obtain an overall impression of the changes. The limitations occur because one must use truncated or simplified potentials to minimise the computer running time and many mechanisms of energy dissipation will be ignored.

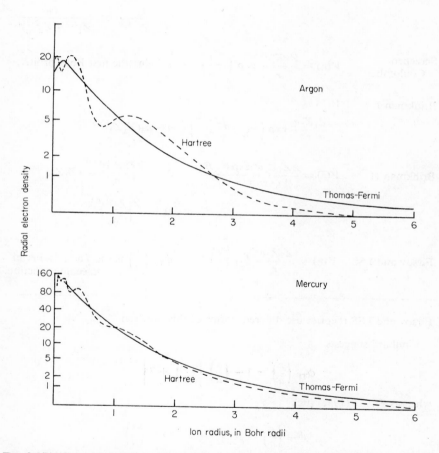

Fig. 2.4(b). Comparisons of the Thomas–Fermi and Hartree interatomic potentials for the ions of argon and mercury.

TABLE 2.1 Examples of interatomic potentials.

Born–Mayer	$V(r) = A \exp\left(-\dfrac{r}{b}\right)$		A, b are constants
Coulomb	$V(r) = \dfrac{Z_1 Z_2 e^2}{r}$		
Neilson	$V(r) = \dfrac{Z_1 Z_2 e^2 a}{r^2} \exp(-1)$		where $a = \dfrac{a_0}{(Z_1 Z_2)^{1/6}}$ or $\dfrac{a_0}{(Z_1^{2/3} + Z_2^{2/3})^{1/2}}$
Screened Coulomb	$V(r) = \dfrac{Z_1 Z_2 e^2}{r} \exp\left(-\dfrac{r}{a}\right)$		a_0 being the first Bohr orbit.
Brinkman I	$V(r) = \dfrac{Z_1 Z_2 e^2}{r} \exp\left(-\dfrac{r}{a}\right)\left(\dfrac{1-r}{2a}\right)$		$a \sim a_0 (Z_1 Z_2)^{-1/6}$
Brinkman II	$V(r) = \dfrac{A Z_1 Z_2 e^2 \exp(-Br)}{1 - \exp(-Ar)}$		$A = \dfrac{0.95 \times 10^{-6}}{a_0} (Z_1 Z_2)^{7/4}$ $B = \dfrac{(Z_1 Z_2)^{1/6}}{1.5 a_0}$
Firsov and LSS	$V(r) = \dfrac{Z_1 Z_2 e^2}{r} \phi_{TF}\left(\dfrac{r}{a}\right)$		$\phi_{TF}\left(\dfrac{r}{a}\right)$ is the Thomas–Fermi screening function.

Firsov and LSS theories use different forms of this function

Lindhard suggests

$$\phi_{TF}\left(\frac{r}{a}\right) = 1 - \left(\frac{r}{a}\right)\left[\left(\frac{r}{a}\right)^2 + 3\right]^{-1/2}$$

whereas Firsov (1958) uses

$$\phi_{TF}\left(\frac{r}{a}\right) = \chi(Z_1^{1/2} + Z_2^{1/2})^{2/3}\left(\frac{r}{a}\right)$$

where the factor χ is tabulated by Gombas (1949).

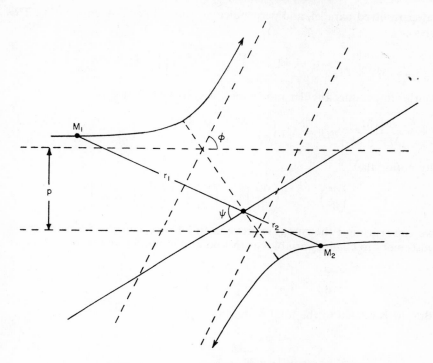

FIG. 2.5. The collision path in centre of mass co-ordinates.

2.4 SCATTERING DURING COLLISIONS

If we now consider ions of a finite size described by a potential function $V(r)$ then we are not limited to collisions where the line between the ion centres is also the direction of motion. Instead, off-axis collisions can be discussed in terms of an impact parameter, p, which is the perpendicular separation of the asymptotes of the hyperbolae which describe the ion trajectories in a centre of mass frame, Fig. 2.5. At any instant the distance of the ions from the centre of mass is

$$r_1 = \frac{M_2 r}{(M_1 + M_2)}, \quad \text{or} \quad r_2 = \frac{M_1 r}{(M_1 + M_2)},$$

where r is the pair separation. For these elastic collisions it is convenient to express the total energy as the sum of potential and kinetic energy with the latter resolved parallel, and perpendicular, to the line between the ions. This gives,

$$\frac{M_2}{(M_1 + M_2)} E_1 = V(r) + \tfrac{1}{2} \frac{M_1 M_2}{(M_1 + M_2)} \left[\left(\frac{dr}{dt}\right)^2 + r^2 \left(\frac{d\psi}{dt}\right)^2 \right].$$

With finite atoms angular momentum is conserved hence

$$\frac{M_1 M_2}{(M_1 + M_2)} u_1 p = M_1 \left(\frac{rM_2}{M_1 + M_2}\right)^2 \frac{d\psi}{dt} + M_2 \left(\frac{rM_1}{M_1 + M_2}\right)^2 \frac{d\psi}{dt}.$$

By noting that

$$\left(\frac{dr}{dt}\right)^2 + r^2 \left(\frac{d\psi}{dt}\right)^2 = \left(\frac{d\psi}{dt}\right)^2 \left[\left(\frac{dr}{d\psi}\right)^2 + r^2\right]$$

we can rearrange the energy expression to be independent of time. It is also customary to change variables at this point and write $u = r^{-1}$ so

$$\frac{du}{d\psi} = \left[\frac{1}{p^2} - \frac{V(u)}{E_1} \frac{(M_1 + M_2)}{M_2 p^2} - u^2\right]^{\frac{1}{2}}.$$

But $d\psi$ is related to the total scattering angle ϕ by

$$\int_{\phi/2}^{\pi/2} d\psi = \tfrac{1}{2} (\pi - \phi)$$

or

$$\phi = \pi - 2p \int_0^{1/u_0} \left[1 - \frac{V(u)}{E_1} \frac{(M_1 + M_2)}{M_2} - p^2 u^2 \right]^{-\frac{1}{2}} du$$

where $1/u_0$ is the distance of closest approach $(R_1 + R_2)$ if we consider an ion "radius".

This equation is important because if we can choose a potential $V(u)$, and integrate, then we can estimate the scattering. To compare the theory with experiment we cannot use the impact parameter so we introduce a differential cross section $d\sigma = 2\pi p \, dp$, which measures the area around the target atom from which a particular scattering angle would occur.

The total cross section for energy transfer

$$E_2 = T = \frac{4 M_1 M_2 E_1}{(M_1 + M_2)^2} \sin^2\left(\frac{\phi}{2}\right)$$

is

$$\sigma = \int_{T_{\min}}^{T_{\max}} \frac{d\sigma}{dE_2} dE_2.$$

2.4 SCATTERING DURING COLLISIONS

Such a cross section equation measures both the energy losses for scattering and the energy transferred which can cause radiation damage. The lower limit of the integral should be chosen in different ways depending on whether we are considering energy loss for the slowing down of a primary or displacement of lattice atoms. In the latter case the lower limit is the threshold energy for removal of a lattice atom from its site.

For a particular scattering angle, the probability that the energy transfer is E_2 is related to the total cross section by

$$P(E_2)\, dE_2 = \frac{1}{\sigma} \frac{d\sigma}{dE_2}\, dE_2.$$

The mathematical problem of choosing a realistic $V(u)$ so that the scattering cross section can be evaluated is quite major. An extreme solution is to consider only glancing collisions where the direction of motion is essentially unchanged. This "impulse" or "momentum" approximation has the interesting result that T is proportional to E_1^{-1} irrespective of $V(u)$. At the other extreme one can use a hard sphere potential i.e.

$$V(r) = \infty \quad \text{for} \quad r < R$$
$$= 0 \quad \text{for} \quad r > R.$$

This gives $d\sigma \propto dT/T_{\max}$.

More realistic potentials can be used if we understand whether we are sensing a close or distant interaction of the ions. For example a Coulomb potential $V(r) = Z_1 Z_2 e^2/r$ leads to $d\sigma \propto E_1 dT/T^2$ but is only appropriate for close penetration of the ions. The screened Coulomb potential used by Bohr (1948), $V(r) = Z_1 Z_2 e^2/r \exp(-r/a_0)$ overcompensates and is too weak at large separations. A range of intermediate examples considered by Lindhard and his co-workers (1961, 1963) were power potentials of the form

$$V(r) \approx \frac{Z_1 Z_2 e^2}{r^s} \frac{a^{s-1}}{s},$$

where $a = 0.885\, a_0 (Z_1^{2/3} + Z_2^{2/3})^{-1/2}$, these gave an elastic cross section

$$d\sigma \approx \frac{C_n}{T_{\max}^{1-1/s}} \frac{dT}{T^{(1+1/s)}} \quad \text{for} \quad s > 1.$$

Here

$$C_n \approx \left(1 - \frac{1}{s}\right) \frac{\pi^2}{2\cdot 718} \frac{e^2 a_0 Z_1 Z_2 M_1}{(Z_1^{2/3} + Z_2^{2/3})^{1/2} (M_1 + M_2)}$$

reflects the underlying use of a modified classical Bohr atom. This cross section is also independent of primary energy in the case of an inverse square law potential, $s = 2$.

In attempts to find an analytic expression one is frequently making intelligent guesses at the form of the screening function. For example Firsov (1958) uses an expression

$$a = 0.885\, a_0\, (Z_1^{1/2} + Z_2^{1/2})^{-2/3}.$$

This differs from the preceding expression for the parameter a by 10 to 20%. Differences of this magnitude are of minor importance in the development of a range theory.

A variation of the inverse square potential is the Nielsen (1956) potential

$$V(r) = \frac{Z_1 Z_2 e^2 a_0}{r^2} \exp(-1).$$

It has the property that it intersects the screened Bohr potential at the value $r = a_0$. The major advantage of this potential is that Nielsen derives from it a quantitative measure of the total ion range, $\bar{R}$, as

$$\bar{R} = 0.6 \frac{(Z_1^{2/3} + Z_2^{2/3})^{1/2}}{Z_1 Z_2} \frac{(M_1 + M_2) M_2}{M_1} \frac{10^{-6} E}{D}$$

where D is the target density and E is measured in keV. Unfortunately it is only applicable for $M_1 > M_2$ and M_1, M_2 between 10 to 200. We shall also see (Section 2.6) that a prediction of the projected range would be preferable to the total range. However, as an easily evaluated first assessment of the range the above expression is useful.

The Lindhard, Scharff and Schiøtt (1963) potential used in most range calculations employs a Thomas–Fermi model of the atom to give

$$V(r) = \frac{Z_1 Z_2 e^2}{r} \phi_{\text{TF}}\left(\frac{r}{a}\right)$$

where the Thomas–Fermi function ϕ_{TF} is a numerical screening function. This has been tabulated by Gombas (1956) and a similar function used by Firsov (1958). Lindhard (1965) offers an approximate analytical form for ϕ_{TF} as

$$1 - \left(\frac{r}{a}\right)\left[\left(\frac{r}{a}\right)^2 + 3\right]^{-1/2}.$$

Alternative power law expressions for the function are used by Winterbon et al. (1970) in the calculation of energy transfer to the target. It is therefore essential to look more closely at the Thomas–Fermi potential and compare it with a more detailed electron density picture obtained by a Hartree or a Hartree–Fock calculation. Figure 2.4(b) shows such a comparison for argon and mercury ions. It is clear that the Thomas–Fermi model represents the same total electronic charge but it has lost the detailed shell structure and will also produce a potential function which is too repulsive at large distances (i.e. > 3 Bohr radii for argon ions). Consequently calculations which use this potential will overestimate the energy loss at large ion separations. Hence ranges predicted by LSS theory are likely to be too short for heavy ions, which are intrinsically large, since the size of the Hartree and Thomas–Fermi ions will deviate most at large Z. Indeed Neilson *et al.* (1973) measure a range which is up to a factor of 2 greater than the predicted LSS range of heavy ions entering aluminium (see Section 2.11).

The LSS cross section for nuclear collisions is valuable because it is appropriate for all values of Z_1, Z_2, M_1, M_2. It is therefore sensible to relate the cross section to dimensionless energy and range parameters ε and ρ,

$$\varepsilon = E \frac{aM_2}{Z_1 Z_2 e^2 (M_1 + M_2)}$$

$$\rho = RNM_2 \frac{4\pi a^2 M_1}{(M_1 + M_2)^2}$$

where R is the range, and N the number of atoms per unit volume.

The stopping cross section, σ, is related to these parameters by

$$\left(\frac{d\varepsilon}{d\rho}\right)_{\text{nuclear}} = \sigma \frac{(M_1 + M_2)}{4\pi e^2 Z_1 Z_2 M_1}.$$

The universal curves computed from this expression are shown in Figs 2.6 and 2.7. In Fig. 2.6 is plotted the differential cross section computed by LSS using a Thomas–Fermi model to give $d\sigma$ in terms of energy transfer as

$$d\sigma = \frac{\pi a^2}{2t^{3/2}} f(t^{1/2}) \, dt$$

since

$$t = \varepsilon^2 \sin^2\left(\frac{\phi}{2}\right) = \varepsilon^2 \left(\frac{T}{T_{\max}}\right).$$

One can see that the LSS and Coulomb potentials predict the same scattering cross sections at high energies where we measure Rutherford scattering. In an intermediate energy range there is also some agreement with the cross section predicted by the inverse square potential.

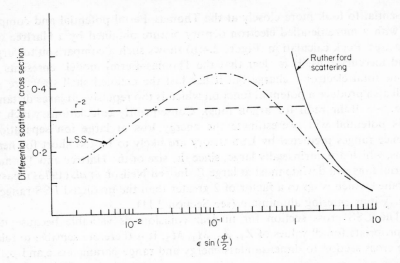

FIG. 2.6. The differential cross section for elastic nuclear collisions. The calculation from the LSS theory using a Thomas–Fermi potential approaches the Rutherford scattering cross section at high energy. For comparison the cross section for an inverse square law potential is also shown. The cross section and energy axes are plotted in terms of universal functions (see text).

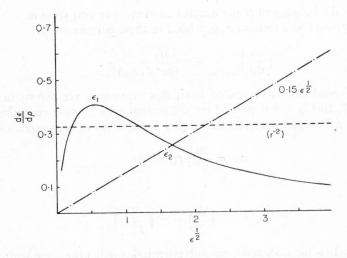

FIG. 2.7. Nuclear and electronic stopping cross sections in reduced units. The curved line is the nuclear cross section computed from the LSS theory of Fig. 2.6. Electronic stopping is proportional to velocity and for comparison the cross section for the inverse square law potential is also shown.

One should note that the differential cross section for nuclear collisions has taken on a general form because the changes produced by scaling the energy (and hence ϕ) are compensated by the $t^{1/2}$ term. An analytic form for $t^{1/2}$ is given by Winterbon et al. (1970, 1972) as

$$f(t^{1/2}) = \lambda t^{(1/2-m)} [1 + (2\lambda t^{(1-m)})^q]^{-1/q}$$

which approaches a power law, $\lambda t^{(1/2-m)}$, for small values of t. The parameters for this power law, λ, m and q are chosen empirically with λ and q being fairly constant, but the important exponent, m, is sensitive to the shape of the tail of the potential (i.e. small values of t).

The presentation of the universal cross section curve in terms of the cross section and velocity of the primary ion is shown in Fig. 2.7. Here the elastic (nuclear) energy losses are compared with the inelastic ones discussed in the next section.

To give some quantitative feel for the energies at which the nuclear cross section is a maximum, ε_1, or the two processes are comparable, ε_2, Tables 2.2 and 2.3 list some representative values of ε/E, ρ/R, ε_1 and ε_2 for a variety of ion target combinations. As a guide line we can equate ε_1 and ε_2 with ε values of 0·35 and 3.

In this section we have outlined how the Lindhard, Scharff, Schiøtt (1963) theory has developed; for a detailed discussion of the validity of this and other potentials the reader is referred to the books by Carter and Colligon (1968) and Torrens (1972), or the conference proceedings edited by Gehlen et al. (1972).

2.5 ELECTRONIC ENERGY LOSS

A classical approach is also appropriate when considering the inelastic energy loss from the passage of the ion through the electronic cloud of the target atom. Again one can trace the historical approach from the early theory of Bohr (1913) where he considered a fully ionised atom striking a second ion. The primary ion will be stripped of all its electrons if it is moving at a higher velocity than the electrons of the K shell ($v > Z_1 e^2/\hbar$) so the rate of energy loss will depend on the closeness of approach to the second atom and thus the number of electrons which can be excited. Bohr wrote the energy loss as

$$\left(\frac{dE}{dx}\right)_{\text{electronic}} = \frac{4\pi Z_1^2 e^4}{mv^2} B$$

where B is a measure of the penetration through the electron shells.

In terms of an impact parameter

$$B = Z_2 \ln\left(\frac{p_{\max}}{p_{\min}}\right),$$

$p_{\min}$ corresponds to the distance of closest approach for head-on collisions. Bethe (1930) obtained a similar expression from a quantum mechanical approach but wrote B in terms of an average excitation energy (I) as

$$B = Z_2 \ln\left(\frac{2mv^2}{I}\right).$$

There followed various attempts to make dE/dx more sophisticated by adding scaling parameters to I in terms of Z_2, plus extra terms to allow for excitation from different electron shells. One should also correct Z_1 since as the primary slows down it will capture electrons and be effectively screened. Despite these inherent difficulties theoreticians have persevered with this problem. The major incentive is now to include complete orbital descriptions so that the theory can be applied to ions slowing down during the process of channelling. Such an approach considered by Cheshire et al. (1969) will be mentioned in Chapter 3.

There is no universal curve which describes the electronic loss for all ion pairs, as there is for elastic collisions, and for amorphous materials the best one can do is to assume the energy loss is proportional to the ion velocity. The Lindhard, Scharff (1961) formulation is

$$\left(\frac{dE}{dx}\right)_{\text{electronic}} = \xi_e \frac{8\pi e^2 N a_0 Z_1 Z_2}{(Z_1^{2/3} + Z_2^{2/3})^{3/2}} \frac{v}{v_0},$$

where v is the velocity, v_0 the Bohr velocity ($Z_1 e^2/\hbar$) and numerically $\xi \sim Z^{1/6}$. When written in terms of ε and ρ the electronic loss is

$$\left(\frac{d\varepsilon}{d\rho}\right)_{\text{electronic}} = k\varepsilon^{1/2}$$

with

$$k \approx \frac{0.079\, Z_1^{1/6} Z_1^{1/2} Z_2^{1/2} (M_1 + M_2)^{3/2}}{(Z_1^{2/3} + Z_2^{2/3})^{3/4} M_1^{3/2} M_2^{1/2}},$$

in other words, a linear energy loss with velocity, and a proportionality constant a function of each ion pair. For implantation experiments the most direct way to use the information is to tabulate k for a particular combination of ion and target atom and then use this to calculate the range ρ for a particular ε. Representative data is included in Tables 2.2. and 2.4. One should note that at higher energies than are typically used for ion implantation the linearity of $d\varepsilon/d\rho$ with $\varepsilon^{1/2}$ is inappropriate as relativistic corrections and nuclear interactions occur. As a consequence the relationship peaks at some point, ε_3, and typical values are seen in Table 2.3.

TABLE 2.2 LSS computed values of ε, ρ and k for ions implanted in Si, Ge and CdTe. The range estimates assume densities of Si = 2·33, Ge = 5·32, CdTe = 5·84 g cm^{-3}.

Z_1	Ion Z_2	ε/E(keV) Si 14	Ge 32	CdTe 50	ρ/R (microns) Si	Ge	CdTe	k Si	Ge	CdTe
3	Li	0·221	0·089	0·052	28·0	8·0	2·7	0·28	0·65	1·04
5	B	0·113	0·049	0·029	32·2	10·6	3·8	0·22	0·47	0·75
7	N	0·074	0·033	0·020	32·2	11·8	4·5	0·20	0·42	0·65
13	Al	0·028	0·015	0·0093	30·5	15·3	6·4	0·14	0·26	0·39
15	P	0·021	0·012	0·0078	29·0	15·7	6·8	0·14	0·24	0·36
31	Ga	0·0054	0·0037	0·0027	17·9	15·2	8·1	0·12	0·16	0·21
33	As	0·0048	0·0034	0·0025	17·0	14·8	8·1	0·12	0·16	0·20
49	In	0·0021	0·0017	0·00130	11·4	12·2	7·6	0·11	0·14	0·17
51	Sb	0·0019	0·0015	0·00121	10·7	11·9	7·5	0·11	0·14	0·16
81	Tl	0·00070	0·00062	0·00052	6·0	8·2	5·9	0·11	0·13	0·14
82	Bi	0·00066	0·00059	0·00050	5·8	8·0	5·8	0·11	0·13	0·14

(Based on the calculated tabulation of Mayer, Eriksson and Davies, 1970)

TABLE 2.3 Energies of the points ε_1 and ε_2

Ion	ε_1 Si	Ge	CdTe	ε_2 Si	Ge	CdTe
B	3	7	12	17	13	10
P	17	29	45	140	140	130
As	73	103	140	800	800	800
Sb	180	230	290	2000	2000	2000
Bi	530	600	700	6000	6000	6000

As a guide $\varepsilon_1 \approx \varepsilon/3$ and $\varepsilon_2 \approx 3\varepsilon$. (Based on the calculated tabulation of Mayer, Eriksson and Davies, 1970).

TABLE 2.4 ρ values as a function of ε and k computed from the LSS theory

ε	$k=0.0$	$k=0.10$	$k=0.12$	$k=0.14$	$k=0.2$	$k=0.3$	$k=0.4$	$k=1.0$
0.01	0.072	0.069	0.069	0.068	0.067	0.064	0.062	0.052
0.02	0.115	1.110	0.109	0.108	0.106	0.102	0.098	0.081
0.05	0.218	0.207	0.205	0.203	0.197	0.188	0.180	0.144
0.10	0.360	0.339	0.335	0.332	0.321	0.304	0.289	0.224
0.20	0.614	0.571	0.563	0.553	0.533	0.501	0.472	0.353
0.50	1.35	1.21	1.19	1.17	1.10	1.01	0.938	0.656
1.0	2.67	2.29	2.22	2.17	2.01	1.80	1.63	1.06
2.0	5.84	4.57	4.39	4.22	3.79	3.26	2.88	1.71
5.0	19.4	11.9	11.1	10.4	8.83	7.11	5.99	3.17
10.0	53.6	23.8	21.6	19.9	16.1	12.3	10.1	4.92

(Based on the calculated tabulation of Mayer, Eriksson and Davies, 1970)

2.6 CHARGE EXCHANGE LOSSES

The cross section for transfer of an electron from one ion to another is a maximum if the ion velocity is close to the orbital velocity of the electron. The process will be well defined for inner shell electrons but as the ion slows down the outer electron shells, particularly for heavy ions, will offer a range of states suitable for charge transfer. For light ions Bohr (1948) proposed capture and loss cross section as

$$\sigma_{capture} \sim 4\pi a_0^2 Z_1^5 Z_2^{1/3} \left(\frac{v_0}{v}\right)^6$$

$$\sigma_{loss} \sim \pi a_0^2 Z_1^{-1} Z_2^{2/3} \left(\frac{v_0}{v}\right)$$

whereas for heavy ions Bohr and Lindhard (1954) suggest

$$\sigma_{capture} \sim \pi a_0^2 Z_1^2 Z_2^{1/3} \left(\frac{v_0}{v}\right)^3$$

$$\sigma_{loss} \sim \pi a_0^2 Z_1^{1/3} Z_2^2 \left(\frac{v_0}{v_1}\right)^3.$$

If we read v_1 as the velocity of outer electrons on the moving ion then for heavy ions the capture and loss cross sections are symmetric about Z_1 and Z_2.

Charge exchange losses represent a small fraction of the total energy losses, typically a few per cent.

2.7 ION RANGES IN AMORPHOUS SOLIDS

The preceding theories indicate the mechanisms of energy loss so we are now able to predict the total path covered by the ion as it is brought to rest from an initial energy E_0. It is

$$R_{\text{total}} = \int_0^{E_0} \frac{-\mathrm{d}E}{(\mathrm{d}E/\mathrm{d}x)_{\text{total}}}.$$

At high energies, where electronic losses dominate, the particle is only slightly deflected but at the end of the path the elastic nuclear collisions produce large angle scattering. The average depth of implant is therefore less than the total range. The more important parameter, the projected range R_P, is the average depth to which an ion will penetrate when the incident beam is normal to the surface. It is related to the total range and the lateral spread, $R_\perp$, of the beam as shown in Fig. 2.8. The early theory of Lindhard and Scharff (1961) provides an approximate relation between the total and projected ranges in the case of nuclear stopping with

$$\frac{R_{\text{total}}}{R_p} \approx 1 + \frac{M_2}{3M_1}.$$

The more detailed LSS theory includes electronic loss and numerical solutions to their equations have been developed by Schiøtt (1966). Various authors have produced tables to describe the ion range, projected range, range straggling (i.e. spread in range) and contributions to the energy loss for different ion-target combinations and energies. These tables by Sigmund

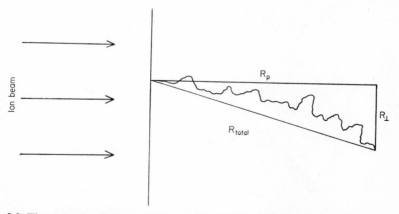

FIG. 2.8. The path of an ion entering a solid showing the total range, the projected range R_P and the lateral spread $R_\perp$.

and Sanders (1967), Channing and Turnbull (1968), Johnson and Gibbons (1970) and Smith (1971) are of fundamental importance in the design of ion implanted devices. For this reason a selection of the range–energy data is provided graphically in Appendix I. For device applications one must consider both the distribution of implanted ions and the energy loss since this may produce atomic displacements. In terms of property changes these can be as important as the implant. The distribution of the radiation damage was considered by Winterbon *et al.* (1970) and will be discussed later. A review of both implantation and damage profiles has been given recently by Sigmund (1972, a, b).

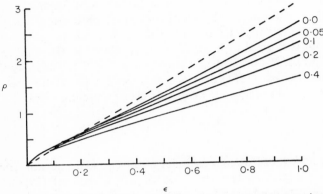

FIG. 2.9. The LSS range–energy curves in reduced units at low energies. The data is shown for various contributions of the electronic energy loss (k values). The dotted line shows the range estimate for an inverse square potential between nuclei.

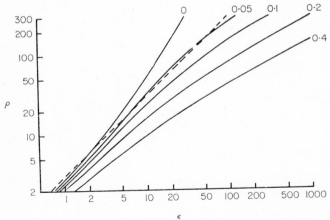

FIG. 2.10. LSS range–energy curves at high energies.

The range energy relations shown in Figs 2.9 and 2.10 demonstrate the importance of electronic loss at high energies for all ion-atom combinations by presenting the data in terms of the reduced parameters ρ and ε. For nuclear scattering the inverse square potential (Nielsen) and the Thomas–Fermi model (LSS) give similar ranges but these diverge according to the strength (k) of the inelastic loss.

At the lowest energies ($\varepsilon \leqslant 0\cdot1$) the ion scarcely penetrates the atom and the Thomas–Fermi function can be approximated by $V(r) \propto r^{-3}$ (Schiøtt, 1970) to provide a range–energy relation

$$\rho \approx \frac{3}{2} \varepsilon^{2/3}$$

which is essentially independent of k.

For applications of the theory it is more appropriate to work in normal units rather than the parameters ρ and ε and we should note that

$$\varepsilon = \frac{32\cdot 5\, M_2 E}{(M_1 + M_2) Z_1 Z_2 (Z_1^{2/3} + Z_2^{2/3})^{1/2}} \qquad (E \text{ in keV})$$

$$\rho = \frac{166\cdot 8 M_1 R}{(M_1 + M_2)^2 (Z_1^{2/3} + Z_2^{2/3})} \qquad (R \text{ in } \mu\text{g cm}^{-2})$$

It is customary to write ranges in units of μg/cm² rather than in units of length since the data can then be compared between different materials without any normalisation. A knowledge of the atomic mass and atomic density gives a simple conversion to length (length = R/density).

It is also useful to write a characteristic energy for a system, E', corresponding to $\varepsilon = 1$. This is given in practical units of keV by

$$E' = \frac{3\cdot 08 \times 10^{-2}\, Z_1 Z_2 (M_1 + M_2)}{(Z_1^{2/3} + Z_2^{2/3})^{1/2}\, M_2}.$$

To estimate the scattering we must compute the standard deviations from the average range and projected range (ΔR and ΔR_P). If the atoms are of similar mass $\Delta R \approx \Delta R_P$ but if $M_2 \gg M_1$ then $\Delta R_P \gg \Delta R$.

At very low energies $\varepsilon \leqslant 0\cdot1$ the mean deviations are simply a function of M_2/M_1 and the random nature of the collisions produces a Gaussian implant profile. The parameters ΔR, ΔR_P, R and R_P are related in a manner shown by Fig. 2.11. In addition it is important to predict the lateral spread of the beam, $R_\perp$, as this sets the limit to the sharpness with which we can

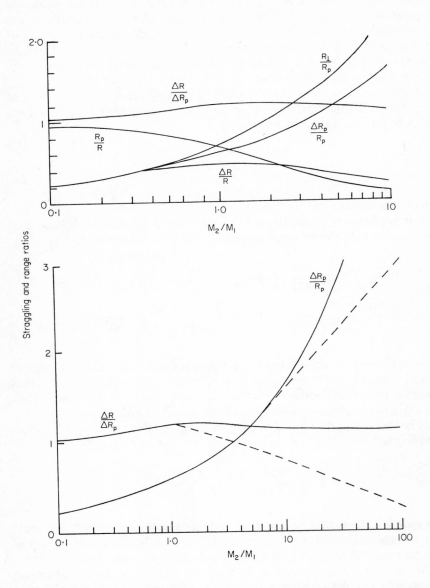

FIG. 2.11(a). The range ratios for low energy ions calculated by Schiøtt (1970) as a function of mass ratio for target and incident ions.

FIG. 2.11(b). The range straggling ratios calculated by Schiøtt (1970) for low energy ions. The dashed curve includes a correction for electronic stopping.

define an implanted region. In this energy range ($\varepsilon \leqslant 0\cdot 1$) the projected range is given by Schiøtt (1970) as

$$\rho_P(\varepsilon) \approx \frac{3}{2} C\left(\frac{M_2}{M_1}\right) \varepsilon^{2/3},$$

or

$$R_P \approx C\left(\frac{M_2}{M_1}\right) M_2 \left(\frac{(Z_1^{1/3} + Z_2^{2/3})E}{Z_1 Z_2}\right)^{2/3}$$

(units of $\mu g\ cm^{-2}$ and keV)

where the function $C(M_2/M_1)$ is approximately a constant (R_P/R_{total}) for $M_2 < 3M_1$ but varies as shown in Fig. 2.12. Electronic losses are important at high mass ratios and the total and projected range stragglings will differ. This is indicated in the right-hand side of Fig. 2.11 where the ratio $\Delta R/\Delta R_P$ has been corrected for inelastic scattering and is shown by a dashed line.

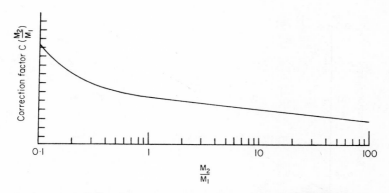

FIG. 2.12. The range correction function $C(M_2/M_1)$ used by Schiøtt (1970).

For intermediate energies $0\cdot 1 \leqslant \varepsilon \leqslant 100$ the Nielsen expression, $\rho \approx 3\cdot 06\ \varepsilon$, can only be used for elastic losses. The range straggling is strongly dependent on k and LSS data, Fig. 2.13, demonstrates how the mean deviation in the range varies with a combination of nuclear and electronic loss. Schiøtt (1970) suggests an approximation in this range of

$$R_P \approx C_1\left(\frac{M_2}{M_1}\right) M_2 \frac{(Z_1^{2/3} + Z_2^{2/3})^{1/2}}{Z_1 Z_2} E. \quad \text{(in } \mu g\ cm^{-2} \text{ and keV)}$$

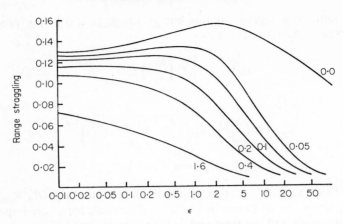

FIG. 2.13. The dependence of the relative straggling in the ion range as a function of reduced energy and electronic energy loss. The data is presented in terms of

$$\frac{(M_1 + M_2)^2}{4 M_1 M_2} \frac{\overline{\Delta R^2}}{\overline{R^2}}.$$

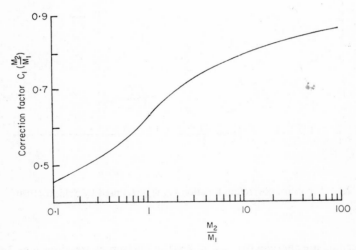

FIG. 2.14. The correction factor $C_1 (M_2/M_1)$ used by Schiøtt (1970) for intermediate energy ions.

Here the function $C_1(M_2/M_1)$ is given by Fig. 2.14 and is useful from $0.5 \, E' < E < 10 \, E'$. Range straggling occurs mainly at the end of the track and $\Delta R_P \sim \Delta R$ and both are a small fraction of the total range. Numerical values for different ion–atom situations are included in Appendix I.

2.8 ION RANGES IN COMPOUNDS

The numerical solutions of the LSS theory presented in data tables are normally intended for studies of ions entering a single element system or compound semiconductors. For wider applications it is necessary to extend these results to other systems. Range scaling in terms of density may be valid for low energy implants since the elastic scattering follows a universal energy loss curve. In practice the range expression of Schiøtt (1970) is also effective even for quite high energies up to 10 E'.

For compound systems estimates of the range and straggling may take the form

$$\frac{1}{R_P} = \sum_i \frac{X_i}{R_{P_i}}$$

and

$$\frac{R^2}{\Delta R^2} = \left[\sum_i \gamma_i \frac{X_i R_i^2}{R_i \Delta R_i^2}\right] \left[\sum_i \gamma_i \frac{X_i}{R_i}\right]^{-1}$$

where

$$\gamma_i = \frac{4 M_1 M_i}{(M_1 + M_i)^2},$$

X_i is the fraction of the total mass of the ith component. Even if the elements involved do not have the same energy dependence for the stopping cross sections these estimates are generally accurate within 10%.

Deviations from the theory are produced by the act of implantation as this changes the density and order of the substrate. In addition the implanted atoms may move by diffusion so an overall prediction which is accurate to 10% is very acceptable.

In some circumstances the ion beam may not be normal to the surface but at an angle θ to the surface normal, the depth straggling, Δx, is then a mixture of range and lateral straggling which is estimated by Schiøtt (1966) as

$$\Delta x^2 = \Delta R_P^2 \cos^2 \theta + \tfrac{1}{2} R_\perp^2 \sin^2 \theta.$$

2.9 CONCENTRATION PROFILES

In amorphous substances the scatter in projected range about the mean projected range is approximately a Gaussian function. If all the implanted ions are contained within this envelope it may be useful to estimate an average concentration, centred at a depth R_P, as

$$\overline{N(x)} = N_D/(2 \cdot 5 \, \Delta R_P),$$

where N_D is the number of implanted ions per unit area.

For a more detailed picture of the implant at a depth x we can write

$$N(x) \approx \frac{N_D}{2 \cdot 5 \Delta R_P} \exp\left(-\frac{(x-R_P)^2}{\Delta R_P{}^2}\right).$$

To achieve a uniform implant concentration with depth it is necessary to make several implantations at different energies to approximate a plateau by the sum of a set of Gaussian distributions. For example the profile shown in Fig. 2.15 of N^+ ions injected into silica successfully produced the characteristics of a plateau when the region acted as an optical waveguide. In practice the ripples on the "plateau" need not exist as they can be smoothed out by diffusion in a subsequent heat treatment or by radiation enhanced diffusion during the bombardment. It should be realised that normal diffusion coefficients are not relevant in the implanted region because of the associated high concentration of vacancies. The effective coefficient is a maximum at

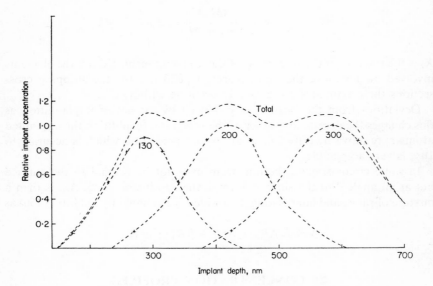

FIG. 2.15. The calculated profile of implanted ions produced by injecting N^+ into silica at 130, 200 and 300 keV in a dose ratio of 61:85:99. Such an implant produced a waveguide for 633 *nm* light.

the position of maximum implant therefore annealing will at first lead to a square topped concentration profile, Fig. 2.16, rather than a general spreading of the implant. Such an effect is ideal for generating a sharp step or a uniform plateau in the implanted impurity concentration. For example Shannon *et al.* (1970) injected phosphorous ions at 100 keV into silicon at room temperature. For an implant dose of 2×10^{16} ions per cm^2 the peak concentration, as judged by the LSS range profile, exceeds the solubility limit for phosphorous in silicon. Therefore annealing at low temperatures, e.g. 650°C, merely flattens the profile to leave a uniform plateau. However, if

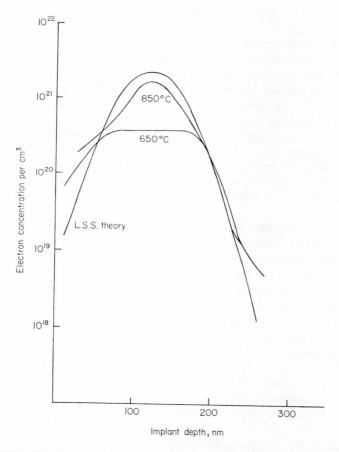

FIG. 2.16. The electrical activity profiles of a phosphorous implant of 2×10^{16} ions cm^{-2} injected at 100 keV into a sample of silicon at room temperature. The original profile is assumed to follow the LSS depth distribution but this is modified by annealing at 650°C or 850°C. At 650°C the phosphorous concentration exceeds the solubility limit of P in Si.

the anneal is made at much higher temperatures, e.g. 850°C, the implanted dose can be accommodated in the silicon as the solubility increases with temperature.

Examples of devices which require steps in the electrical properties are microwave devices such as the n–p–n device made by successive As and B implants in silicon (Barnoski and Loper, 1973).

Alternative procedures to obtain uniform implants can include implantation with a steadily changing acceleration potential or rotation of the specimen to the beam direction.

2.10 IMPLANTATION MASKS

Many of the advantages of ion beam technology derive from the directional nature of the implant and the ability to define an implant pattern by means of a mask. The masking materials can also serve secondary functions of passivating the surface during implantation or inhibiting the build up of charge in the case of insulators. A complete coating of the surface inhibits sputtering of the substrate atoms and in compound materials prevents selective loss of the components if the compound is unstable in vacuum at the working temperature. The mask density and thickness must be noted when estimating the implant profile. Such corrections are necessary because a layer of a heavy element, say 100 Å of silver or gold, is equivalent to several hundred angstroms of substrate material in the case of a light compound such as silica.

If the mask is to define the beam then one must also calculate the thickness of mask material which will completely attenuate the beam. Photoresist materials used in the semiconductor industry are basically carbon compounds in which ion ranges are quite long. An added advantage of a thick mask is that photoresists polymerise during ion bombardment and are then chemically less soluble. However, if the layer is thick the solvents attack the photoresist at the interface with the substrate where polymerisation had not occurred, so the mask is detachable.

Implantation through a masking layer has several side effects. Any sputtering which occurs can take place from the mask rather than the sample but atoms from the mask will be driven into the substrate by direct collisions and radiation enhanced diffusion. This produces a better bond between the two layers. If the purposes of the mask was to provide a temporary protection or a conducting layer this enhanced bonding is a nuisance, but in other circumstances the enhanced adhesion can be an asset.

In extreme cases of heavy implant ions, short ranges, no coatings and high doses, the surface will sputter at a high rate. If the surface recedes during implantation there is a maximum implant concentration which itself will act as a masking layer and the simple LSS theory must be modified.

Mask preparation is an art in its own right and the reader is referred to such articles as those of Stephen (Dearnaley *et al.* 1973), or Section 9.7 of this book.

Chemical preparation of photoresists does not always produce ideal edge definition of the implant region because of the inherent grain size of the photographic images and the undercutting effects of the chemical development of the unpolymerised material.

Some improvements are possible by defining the masking pattern with a fine electron or ion beam, either by directly sputtering a mask or altering its stability to chemical attack. The technique used will vary with the problems of a particular device.

2.11 EXPERIMENTAL MEASUREMENT OF ION RANGES

The theories mentioned so far are applicable to amorphous materials whereas many substances of interest are polycrystalline or single crystals. Thus we expect a depth profile for the implanted ions which is a mixture of the LSS range and a longer range component corresponding to ions which were channelled for some part of their path. Ideally these might combine to give a range profile as shown in Fig. 2.17. To determine experimentally the

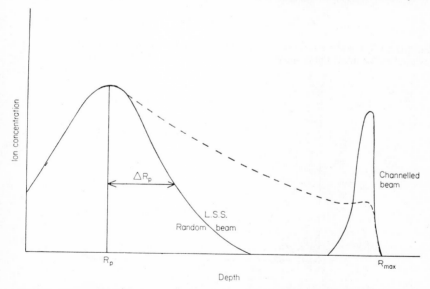

FIG. 2.17. The depth distribution of implanted ions in single crystals. The two peaks arise from normal LSS range considerations and a deep peak from channelled ions. The region between the two will vary in concentration depending on the rate of dechannelling.

profile it is important to realise that we must detect specific atoms (M_1) and not measure a change in property, since many chemical, electrical and optical changes result from both implantation and radiation damage. We must therefore label the implant atoms by either using radioactive isotopes or use a nuclear reaction. Alternatively we may sense a characteristic property of the atoms such as X-ray emission or an Auger transition. Yet again, if we probe through the region by Rutherford backscattering or surface ion

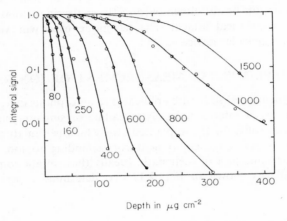

FIG. 2.18(a). Integral penetration distribution for ^{85}Kr into amorphous Al_2O_3. This data was used to construct the range profiles of Fig. 2.18(b). (1μg cm^{-2} ≈ 30 Å).

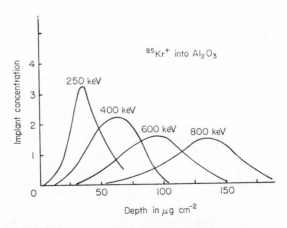

FIG. 2.18(b). Range distributions of ^{85}Kr ions implanted into amorphous Al_2O_3 (after Jespersgaard and Davies, 1967).

mass spectrometry etc. we may be able to identify the implant profile as distinct from the effect of the implant.

The most common method has been to use radioactive implants and measure the residual activity as various layers are stripped from the surface (e.g. Davies *et al.* 1963). A mechanical technique of vibratory polishing with an abrasive slurry has been developed by Whitton (1965, 1968) which is sensitive enough to remove 50 Å layers. With this approach he and later workers have studied ion ranges in Al, Au, Cu, GaAs, Si, Ta, Ta_2O_5, UO_2, W and ZrO_2.

Chemical etching or ion beam polishing provide alternative methods. In such experiments it is essential to calibrate the actual depth change by weight loss or optical interferometry since the stability of the surface is altered by the implantation.

A different approach is feasible for materials which form anodic oxides (e.g. Al, Si, Ta, W) as the oxide layer can be implanted and progressively removed by electro-chemical means to reveal the profile. An example of such results are shown in Fig. 2.18(a). These oxides have been studied in some detail as they are truly amorphous and the classic results of Jespersgaard and Davies (1967) are shown in Fig. 2.18(b) for the penetration of ^{85}Kr into amorphous Al_2O_3. (Note that in Al_2O_3 a penetration of 1 µg/cm² corresponds

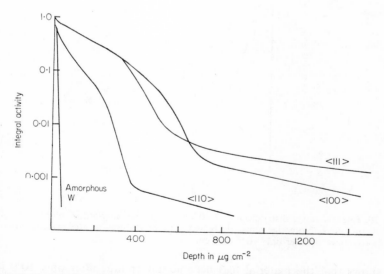

FIG. 2.19. Range information obtained by measurement of the residual radio-activity in a single crystal of tungsten bombarded with 40 keV ions of ^{125}Xe. Note the enhanced ranges if the beam is parallel to a crystal axis compared with the range in amorphous tungsten. (1 µg cm⁻² corresponds to a depth of 5·2 Å).

to a range of 30 Å.) For our purposes the most important result of this type of experiment is that it validates the LSS theory.

Rather than present a range profile many experiments are described in terms of the fraction of the radioactive material which remains beyond a particular depth (e.g. Fig. 2.19). Whilst this can be readily converted to the familiar Gaussian profiles of LSS theory this presentation has the advantage that it emphasises the effects of channelling. It also clearly displays the reduction in channelling which occurs as the channels become blocked by interstitial atoms or radiation damage in high dose experiments. Figure 2.19 is taken from the results of Domeij *et al.* (1964) on the range of radioactive ^{125}Xe injected at 40 keV into tungsten. Extremely long ion ranges are apparent when the ion beam is aligned with crystalline directions, for comparison the penetration in amorphous tungsten is also shown. Some substances show a marked loss of structure when subject to radiation damage and the channelling effects are soon lost at relatively low implant concentrations. Data taken by Whitton *et al.* (1970) for GaAs is shown in Fig. 2.20.

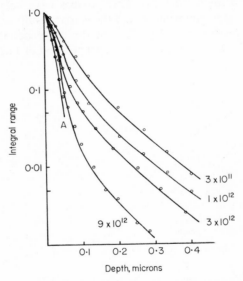

FIG. 2.20. Integral range distributions of 40 keV ^{85}Kr ions implanted along the $\langle 111 \rangle$ direction of GaAs. Note the range decreases with high doses as the material becomes amorphous, curve *A*, after only 10^{13} ions cm^{-2}.

This shows that the material had become amorphous after only 10^{13} ions cm^{-2} under the conditions used in this experiment.

As we mentioned earlier it is possible to simulate implantation with a computer calculation of atomic movements in a model of the solid. In most situations a repulsive pair potential is chosen and the calculations compared with existing experimental data. The situation was reversed by Robinson and Oen (1933) who were computing ion ranges in an ordered crystal lattice. They found that extensive ranges were possible if the ion beam was aligned parallel to crystalling axes and thereby revealed the phenomenon of channelling. It is interesting to note that a similar suggestion made by Stark in 1912 had been forgotten.

2.12 RECENT MEASUREMENTS OF ION RANGES

With the availability of more versatile ion sources many more ion-target range measurements have been made. Not surprisingly it was found that when $M_1 \gg M_2$ there was some divergence in the measured ranges away from the simple LSS ranges. For example in the work of Nielson *et al.* (1973) the ranges are almost twice the LSS range and show a definite Z_1 dependence. (Table 2.5).

TABLE 2.5 Ranges of 100 keV ions in aluminium ($Z_2 = 13$)

Implanted ion	Cs	Ba	La	Sm	Eu	Tb	Au
Atomic number, Z_1	55	56	57	62	63	65	79
Observed range / LSS range	1·81	1·91	2·02	1·36	1·15	1·89	2·19

As was mentioned in Section 2.4 the Thomas–Fermi potential appears to be too repulsive for large ions at large distances and has none of the electronic shell structure that one would expect in a real ion. It would appear that the more recent data is sensing this weakness. However Thompson and Neilson (1974) suspect that they are also detecting a further effect which they attribute to the excitation of inner-shell electrons which transfer energy by an energy level matching process during the collision (see also Neilson *et al*, 1976).

2.13 ENERGY DEPOSITION AND DEFECT FORMATION

In order to specify an ion range we considered the rate at which energy was transferred to the target. If we now consider the recoiling atom and the atomic displacements which it produces as the energy is dissipated we will have calculated a depth profile for the radiation damage. To be more precise we will actually have calculated the depth profile of energy deposition since the detailed structure of the radiation damage will depend on many other parameters, such as the stability of vacancies and interstitials. These details will be discussed in Chapter 4.

To simplify the calculation of the spatial distribution of deposited energy it is customary to assume that electronic energy loss does not contribute to displacements and that some fraction of the recoil energy is also lost in electronic processes.

One approach to the problem has been to use a Monte-Carlo programme in a computer simulation of the damage. A random array of lattice points is generated and after each collision the direction of motion is selected by the programme. A fraction of the transferred energy is also "removed" to simulate electronic energy loss. In this way an entire collision cascade can be constructed. There are limitations to the method because one must assume some mean free path and the depth profile for the energy transfer from the implanted ion (a Gaussian function is often used, e.g. Brice, 1970). Further, the partitioning of energy between elastic and electronic processes is subject to statistical fluctuations so a constant division of the energy will produce a modified damage distribution.

In single crystals where there is an ordered structure the approach may also ignore the transport of energy by replacement collision sequences (see Chapter 4) or channelling. Some of these difficulties are minor and computer

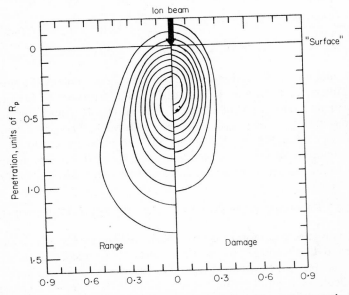

FIG. 2.21. Contours of equal density and equal range calculated for an ion entering a solid ($M_1 = M_2$). Scales are in units of the projected range.

models of damage distributions have been made by Oen *et al.* (1963, 1964) and Pavlov *et al.* (1967).

An alternative approach (Winterbon *et al.*, 1970) is to relate the energy deposition, range and spread in range by a set of integral equations. These were then solved by computing averages of the distribution functions and the moments of these functions. For our purposes the most instructive presentation of the results is given in terms of a contour map of damage (energy deposition) and implant density as a function of depth and lateral spread, Fig. 2.21 (see also Brice, 1970). It is immediately obvious that the damage profile peaks nearer the surface than the implant profile. In this particular example of equal mass for the implanted and target ions the lateral spread of the damage is less than the spread of the implanted ions.

For computational purposes the "surface" is drawn within a continuous medium so contour lines extending beyond the surface should be interpreted with care. In fact they have physical significance and relate to sputtering of atoms from the surface (see Chapter 6).

We already know that for elastic collisions the spread in ion ranges is a function of M_2/M_1 so it is not surprising that there is a corresponding effect in the rate of energy deposition. Figure 2.22 presents the ion range and damage profiles for two cases of $M_2 = 4M_1$ and $4M_2 = M_1$. The detailed shape of these curves is sensitive to the exponent used in the power law

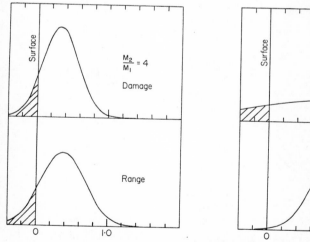

FIG. 2.22. Examples of computed range and damage profiles for target to ion mass ratios of 4 and $\frac{1}{4}$. The vertical lines represent the "surface" in the computer model so the shaded regions may be interpreted as ion reflection or sputtering.

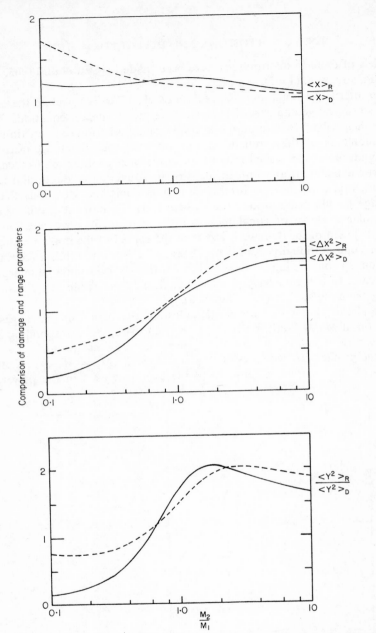

FIG. 2.23. A comparison of range and damage parameters as a function of M_2/M_1. Two exponents of the power law have been considered ---- $m = \frac{1}{3}$; ———— $m = \frac{1}{2}$. (a) First order averages. $\langle X \rangle_R$ = average projected range; $\langle X \rangle_D$ = average damage depth. (b) Second order averages. $\langle \Delta X^2 \rangle_R$ = straggling of projected range; $\langle \Delta X^2 \rangle_D$ = width of the damage depth distribution. (c) Second order averages. $\langle Y^2 \rangle_R$ = transverse straggling of range distribution; $\langle Y^2 \rangle_D$ = transverse width of damage distribution.

approximation of the Thomas–Fermi potential but the general features are maintained. Namely, if $M_2 > M_1$ the distributions are skew and peak near the surface but if $M_2 < M_1$ the implant profile is Gaussian and well defined whereas the damage profile is more uniform with depth.

These trends are displayed more fully in Fig. 2.23 which compares (i) the average projected range to average depth of damage, (ii) the straggling of the projected range with the width of the damage distribution and (iii) the transverse width of the damage region. The figure shows the results of two computations which utilised different exponents in the Thomas–Fermi power law approximation. It is apparent that the approximations yield different results when there is a large difference between the two ion masses.

2.14 CONCLUSION

In the process of slowing down in a solid an implanted ion undergoes many scattering events and provides energy to the solid which may result in atomic displacements. It appears that the mechanisms of energy loss are sufficiently well understood that we can predict both the implant and the damage distributions in the solid with reasonable accuracy. It would be unrealistic to expect general theories to completely satisfy all ion–solid systems for all energies of implantation but the Lindhard, Scharff, Schiøtt theory of ion ranges and the Winterbon, Sigmund, Sanders theory of damage production offer a very real approximation to the measured quantities.

REFERENCES

Barnoski, M. K. and Loper, D. D. (1973). *Sol. Stat. Electr.* **16**, 441.
Bethe, H. A. (1930). *Ann. Physik.* **5**, 325.
Bohr, N. (1913). *Phil. Mag.* **25**, 10.
Bohr, N. (1948). *Kgl. Danske Vid. Selsk. Mat-Fys. Medd.* **18**, No. 8.
Bohr, N. and Lindhard, J. (1954). *Kgl. Danske Vid. Selsk Mat-Fys. Medd.* **28**, No. 7.
Brice, D. K. (1970). *Rad. Eff.* **6**, 77.
Brice, D. K. (1971). *Rad. Eff.* **11**, 227.
Carter, G. and Colligon, J. S. (1968). "Ion Bombardment of Solids", Heinemann.
Channing, D. A. and Turnbull, J. A. (1968). C.E.G.B. Berkeley Rept. RD/B/N 1114.
Cheshire, I., Dearnaley, G. and Poate, J. M. (1969). *Proc. Roy. Soc.* **A311**, 47.
Davies, J. A., Domeij, B. and Uhler, J. (1963). *Arkiv. f. Fysik.* **24**, 377.
Dearnaley, G., Freeman, J. H., Nelson, R. S. and Stephen, J. (1973). "Ion Implantation", North Holland.
Domeij, B., Brown, F., Davies, J. A., Piercy, G. R. and Kornelsen, E. V. (1964). *Phys. Rev. Letts.* **12**, 363.
Firsov, O. B. (1958). *Sov. Phys. J.E.T.P.* **7**, 308.
Firsov, O. B. (1958). Translation *Sov. Phys. J.E.T.P.* **6**, 534.

Gehlen, P. C., Beeler, J. R. and Jaffee, R. I. (Eds) (1972). "Interatomic Potentials and the Simulation of Lattice Defects", Plenum Press.
Gombas, P. (1949). "Die Statistiche Theorie des Atoms und Ihre Anwendungen", Springer-Verlag.
Gombas, P. (1956). *Handbuch der Physik*, **36**, 109.
Jespersgaard, P. and Davies, J. A. (1967). *Can. J. Phys.* **45**, 2983.
Johnson, W. S. and Gibbons, J. F. (1970). "Projected Range Statistics in Semiconductors", Stanford University Bookstore, *ibid* second edition, distributed by Halstead Press (Wiley).
Kessel, Q. C. (1970). "Case Studies on Atomic Collisions I" p. 400, (Eds McDaniel and McDowell), North Holland.
Lindhard, J. (1965). *Kgl. Danske Vid. Selsk. Matt-Fys. Medd.* **34**, No. 14.
Lindhard, J. and Scharff, M. (1961). *Phys. Rev.* **124**, 128.
Lindhard, J., Scharff, M. and Schiøtt, H. E. (1963). *Kgl. Danske Vid. Selsk. Matt-Fys. Medd.* **33**, No. 14.
Mayer, J. W., Eriksson, L. and Davies, J. A. (1970). "Ion Implantation in Semiconductors", Academic Press.
Neilson, G. W., Farmery, B. W. and Thompson, M. W. (1973). *Phys. Letts.* **46A**, 45.
Neilson, G. W., Farmery, B. W., Goode, P. and Thompson, M. W. (1976). *Proc. Surf. Sci.*, June issue.
Nielsen, K. O. (1956). "Electromagnetically Enriched Isotopes and Mass Spectrometry", Academic Press.
Oen, O. S., Holmes, D. K. and Robinson, M. T. (1963). *J. Appl. Phys.* **34**, 302.
Oen, O. S. and Robinson, M. T. (1964). *J. Appl. Phys.* **35**, 2515.
Pavlov, P. V., Tetel'baum, D. I., Zorin, E. I. and Alekseev, V. I. (1967). Translation *Sov. Phys-Sol. State*, **8**, 2141.
Robinson, M. T. and Oen, O. S. (1963). *Phys. Rev.* **132**, 2385.
Rutherford, E. (1911). *Phil. Mag.* **21**, 669.
Schiøtt, H. E. (1966). *Kgl. Danske Vid. Selsk. Matt-Fys. Medd.* **35**, No. 9.
Schiøtt, H. E. (1970). *Rad. Eff.* **6**, 107.
Shannon, J. M., Ford, R. A. and Gard, G. A. (1970). *Rad Eff.* **6**, 217.
Sigmund, P. (1972a). "Physics of Ionized Gases 1972", p. 137 (Ed. M. V. Kurepa), Publisher I. of P. Beograd.
Sigmund, P. (1972b). *Rev. Roum Phys.* **17**, 823, 969, 1079.
Sigmund, P. and Sanders, J. B. (1967). Proc. Intern. Conf. Appl. Ion Beam Semiconductor Techn. (Ed. P. Glotin), Ophrys, Grenoble.
Smith, B. J. (1971). AERE Rept. R6660.
Stark, J. (1912). *Phys. Zeits* **13**, 973.
Thompson, M. W. and Neilson, G. W. (1974). *Phys. Lett.* **49A**, 151.
Torrens, I. M. (1972). "Interatomic Potentials", Academic Press.
Whitton, J. L. (1965). *J. Appl. Phys.* **36**, 3917.
Whitton, J. L. (1968). *Can. J. Phys.* **46**, 581.
Whitton, J. L., Carter, G., Freeman, J. H. and Gard, G. A. (1970). *J. Mat. Sci.* **4**, 208.
Wilson, R. G. and Brewer, G. R. (1973). "Ion beams", Wiley.
Winterbon, K. B. (1972). *Rad. Eff.* **13**, 215.
Winterbon, K. B., Sigmund, P. and Sanders, J. B. (1970). *Kgl. Danske Vid. Selsk. Mat-Fys. Medd.* **37**, 14.

CHAPTER 3

Ion Channelling

3.1 INTRODUCTION

The range–energy considerations of Chapter 2, the production of radiation damage (Chapter 4), sputtering (Chapter 6) or reflection of ions from surface layers (Chapter 8) are most readily discussed in terms of amorphous solids. In practice ion implantation may involve single or polycrystalline targets so we must also consider the penetration of ions into an ordered lattice. The major difference between an amorphous and an ordered array of target atoms is that some ions will suffer a reduced rate of energy loss for that part of their path which is parallel to rows or planes of the crystal. The relative importance of nuclear and electronic energy losses will also differ for these ions compared with the amorphous structure. As a first step the crystal may be thought of as having transparent directions into which the ions may enter. Whilst transparency will enhance the range of a few ions we should realise that the rows and planes of lattice atoms can provide potential walls to steer ions by small angles from a crystal axis. This steering action is termed channelling. Channelled ions deposit their energy deeper into the solid than would randomly directed ions and in addition the steering action precludes direct nuclear events. Nuclear reactions still occur with interstitial ions or at sites where defects distort the perfect lattice so by a comparison of reaction yields for ions directed in a variety of channels it is possible to study the lattice location of the defects in the lattice. The crystallographic aspects of channelling were essentially proposed by Stark (1912) but effectively the subject only developed in the 1960's. In transmission patterns produced by ions channelled through thin films one sees a pattern of spots and lines which readily yield the symmetry and structure of simple lattices. In fact Stark's suggestion was far more direct than the X-ray studies because the channelled pattern offers a direct image of the lattice rather than a reciprocal lattice image. Many ions and crystals have been studied by channelling methods and the subject is well understood and extensively documented. For a detailed view of the subject the reader might proceed to the books and review papers of Lindhard (1965), Tulinov (1965), Datz *et al.* (1967), Thompson (1968), Morgan (1973) and Gemmell (1974).

The study of channelled ions can offer information of the mechanisms of energy transfer between the primary ion and the lattice; be used to test models of interatomic potentials; provide data for lattice location studies; or provide a means of introducing low-loss long-range paths for ion implanted devices. Our major aim in this present book is to examine the prospects for ion implantation in technology beyond the existing area of semiconductors. From this viewpoint the importance of channelling is reduced since many implantation facilities use beams with an angular divergence greater than those required for channelling and when single crystals are targets these are consciously oriented away from channelling directions. The fraction of channelled ions is thus relatively minor. Further, the beam introduces disorder into the lattice, which in turn reduces the ability of the lattice to steer the ions. This was already shown in Fig. 2.20 in which the range profiles of ^{85}Kr ions in a GaAs lattice were markedly reduced for irradiations in excess of 10^{13} ions cm^{-2}. Consequently channelling may not have commercial applications unless one is modifying a property which is sensitive

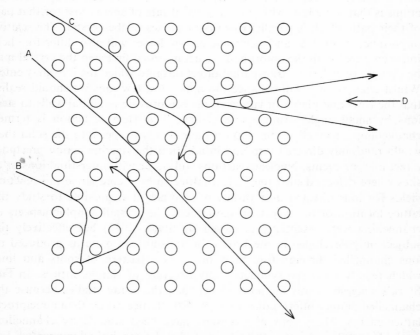

FIG. 3.1. Examples of ion trajectories in a crystal lattice. Path A is a channelling direction, B is random, C is quasi-channelled, D shows a blocking direction for emitted particles.

to small impurity levels (e.g. luminescence or semiconducting materials) and is probably less relevant in the manufacture of metallic alloys where one is usually adding ions in percentage concentrations.

Channelling is equally effective for ions entering the crystal or ions originating within the solid which are approaching the surface. For example a radioactive ion which emits alpha particles will appear to have a yield pattern related to the crystal structure as some directions will block and scatter the particles whilst others can provide channelling. There will probably be differences in the apparent energy spectra of the emitted ions. Figure 3.1 shows examples of random, channelling and blocking directions for a two dimensional array of atoms.

Nuclear reactions can be used as a sensitive detector of dechannelling events. The primary beam must be directed within a small fraction of a degree for channelling to occur and the dechannelling event allows the ion to have a direct nuclear collision and provide a characteristic emission. Some such reactions merely require a threshold energy for penetration whereas others are critically energy dependent for a resonance process. For example a measure of the gamma yield from the resonance reaction of protons in silicon $Si(p, \gamma)$ can provide the energy loss in the channel as well as the range if one uses a variety of sample thicknesses and primary energies. In this example the resonance is peaked at 414 keV, (Bøgh et al., 1964).

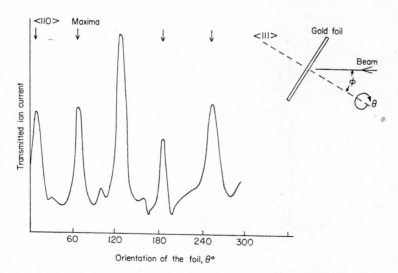

FIG. 3.2. The Nelson and Thompson (1963) experiment to detect channel directions. A specimen, inclined to the beam, is rotated and the transmitted ion current recorded. The gold foil was twinned so shows extra maxima.

Transmission experiments are feasible because of the reduction in energy losses from nuclear collisions and ion ranges can be a factor of ten greater than that of similar energy ions in amorphous solids. Quantitative statements of ion range and the rate of dechannelling may vary because of the difficulties in sample preparation but the general characteristics of the angular dependence of the transmitted or reflected fraction are shown in Figs 3.2 and 3.3.

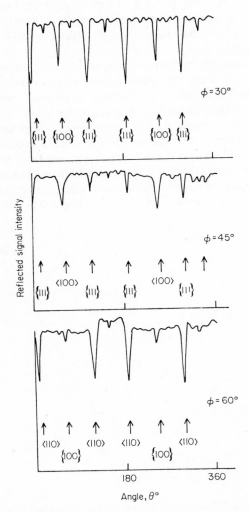

FIG. 3.3. As Fig. 3.2 but a measure of the percentage of primary ions which are backscattered. Note both axial and planar channelling.

These show results from Nelson and Thompson (1963) obtained by rotating a gold foil, inclined to the incident beam, scanned through the crystal planes and axes. (The sample was twinned so exhibits two sets of channelling directions.) By changing the inclination of the foil and beam one changes the relative importance of the different directions and the reflection efficiency but the major directions are still apparent.

3.2 SIMPLE CHANNELLING THEORY

The angular dependence of channelling events provides us with a measure of the reflected yield and an angular channel width, where we may define quantities as in Fig. 3.4. The yield is normalized relative to the random direction yield of an amorphous solid rather than to the maximum value since the high shoulders are also a consequence of an ordered lattice. In

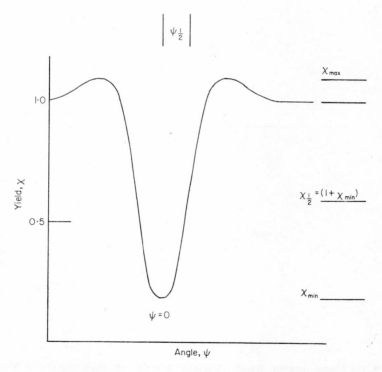

FIG. 3.4. A schematic of the yield as a function of angle across the channel.

practice it is sometimes difficult to obtain the "random" direction. Further information may be obtainable from the energy spectra of the ions and additional nuclear reactions.

One must now develop the theories of channelling events to compare with the experimental data. As with the calculations of ion ranges a computer simulation of the events is informative as one can vary the interatomic potentials and include thermal or impurity effects whilst following individual ion trajectories. Indeed the Robinson and Oen (1963) calculations of ion ranges in copper predicted the experimental results. Nevertheless it is convenient to describe the process in analytical terms as was first done by Lehmann and Leibfried (1963) or Nelson and Thompson (1963). If the particle is to remain channelled then the interaction with any one ion must represent a very minor change in direction and for simplicity we can imagine the lattice atoms are immobile.

If we first consider a two body event as an ion passes a lattice atom then the directional change can be described as a result of a momentum pulse. Figure 3.5 shows that this results from a sideways force produced by the

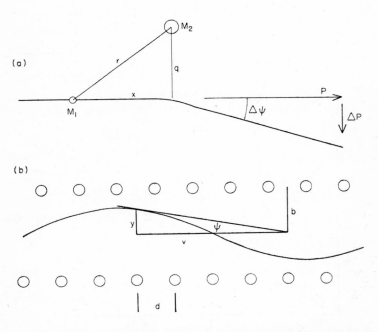

FIG. 3.5. (a) The momentum approximation for a single low angle scattering event. (b) A series of small angle scattering events which steer an ion along an axis.

potential $V(r)$ acting for a time dt. At a velocity v_1, $dt = dx/v_1$, the total change in momentum resolved normal to the x direction is

$$\Delta P = \int_{-\infty}^{\infty} -\frac{dV(r)}{dr} \frac{q}{r} \frac{dx}{v_1}$$

or rewritten in terms of an impact parameter ρ

$$\Delta P = \int_{\rho}^{\infty} -\frac{2\rho}{v_1 (r^2 - \rho^2)^{1/2}} dr \left(= \frac{2}{v_1} I(\rho)\right).$$

The angular change relates the original and new momentum components by

$$\Delta \psi = \frac{\Delta P}{P} = \frac{1}{E_1} \int_{\rho}^{\infty} -\frac{dV(r)}{dr} \frac{\rho \, dr}{(r^2 - \rho^2)^{1/2}}.$$

For axial channelling the ion trajectory is confined by a series of small angle collisions by the rows of lattice atoms. Each event can be considered by the impulse approximation so for successive collisions we may measure the angular rate of change as

$$\dot{\psi} = -\frac{2}{dM_1 v_1} I(b - y)$$

where we measure the y axis from the centre line of the channel. Clearly $\rho = b - y$ and $\dot{\rho} = -\dot{y}$ and the angular change is $\psi = \dot{y}/v_1$ so $\dot{\psi} = \ddot{y}/v_1$. Hence we can write an equation of motion for the particle moving in the y direction in the form

$$M_1 \ddot{y} = -\frac{2}{d} I(b - y),$$

so long as the momentum approximation is appropriate. In so doing we can average the effects of individual collisions and write a continuous potential to replace the string of atoms.

The axial channel potential is

$$U(y) = \frac{2}{d} \int_0^{\hat{y}} I(b - y) \, dy.$$

We derived an axial channel potential because we considered a two dimensional lattice with two rows of atoms. However a brief inspection of a

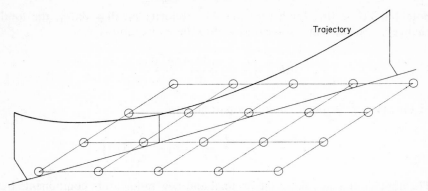

Fig. 3.6. An ion trajectory between crystal planes. Only the lower plane is indicated.

three dimensional picture (Fig. 3.6) shows we might also expect a particle to be confined within planes. However the planar interaction is much weaker than the axial one hence the steering ability is reduced and the critical angle at which dechannelling will occur limits the range of angles accepted for channelling to about a quarter the beam divergence acceptable for planar channelling ($\psi_{1/2} \sim 1°$ for axial or $0.25°$ for planar channels).

To express the channel potential $U(y)$ we must select an interatomic potential $V(r)$ and as was apparent in Chapter 2, there are many alternative analytical forms. The appearance of the potential will probably resemble the examples in Fig. 3.7 calculated by Nelson and Thompson (1963) for ions

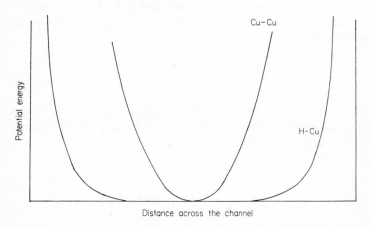

Fig. 3.7. Examples of channel potentials calculated for copper demonstrating the different penetration of the potential into the channel for light or heavy ions (after Nelson and Thompson, 1963).

travelling along the [110] axis in copper. This has the general features that for light ions the wall potential is steep so rapidly falls to a minimum and the central region of the channel is essentially field free, whereas for large ions the repulsive field extends into the centre of the channel, in this case providing a parabolic potential well. For this calculation Nelson and Thompson (1963) used a screened Coulomb potential of the form

$$V(r) = \frac{Z_1 Z_2}{r} e^2 \exp\left(-\frac{r}{a}\right)$$

which led to

$$U(y) = \frac{2(2\pi)^{1/2} Z_1 Z_2 a_0 E_R}{d} \frac{\exp-(b-y)/a}{(b-y)^{1/2} a^{-1/2}}$$

(E_R is the Rydberg energy, 13.6 eV, a_0 is the Bohr radius).

Alternative interatomic potentials lead to channel potentials such as:

$$U(y) = \frac{2 Z_1 Z_2 e^2}{d} H\left(\frac{y}{a}\right)$$

for the Bohr potential used by Lehmann and Leibfried (1963). They also used a Born–Mayer potential to give

$$U(y) = \frac{2 A_{BM}}{d} H(B_{BM} y).$$

(The function $H(y/a)$, $H(B_{BM} y)$ are explained in the original papers.)

The Molière potential used by Erginsoy (1965) contains a variety of correction terms with

$$U(y) = \frac{2 Z_1 Z_2 e^2}{d} \left[0.1 H\left(\frac{6y}{a}\right) + 0.55 H\left(\frac{1.2y}{a}\right) + 0.35 H\left(\frac{0.3y}{a}\right) \right].$$

The Molière potential is very similar in shape to the Lindhard (1965) potential which is often referred to as the "standard continuum potential" with the form

$$U(y) = \frac{2 Z_1 Z_2 e^2}{d} G\left(\frac{y}{a}\right)$$

where the function $G(y/a)$ is approximately

$$\ln\left[\left(\frac{\sqrt{3}a}{y}\right)^2 + 1\right].$$

(Van Vliet (1973) uses a function of half this value.)

3.3 MOTION IN A CHANNEL

Having replaced the atomic structure of a row by a continuum potential we may write the Hamiltonian for the total energy of the projectile in terms of $U(y)$ and kinetic energy. For channelled particles which satisfy the momentum approximation the axial kinetic energy is essentially constant, thus energy transfer can only occur between the potential and transverse energy.

$$E_\perp = U(y) + (p_x^2 + p_y^2)/2M_1.$$

In terms of the angle ψ at which the particle is inclined to the axis

$$E_\perp = U(y) + E\psi^2.$$

The equation indicates an upper limit, ψ_c, for channelling to continue and defines a minimum distance of approach to the wall when $E_\perp = U(\rho_0)$. Thermal effects may be included in the potential by redefining the critical approach distance, for example by an equation of the form

$$\rho_c^2 = \rho_0^2 + k\chi_{rms}^2$$

in which k is a constant ($\sim 1\cdot 5$), ρ_0 is the static value and χ_{rms} is the average thermal displacement of a lattice ion.

Once equilibrium has been achieved within the channel Lindhard (1965) suggested there is an equal probability of finding an ion anywhere in the allowed region of the potential. One can then calculate any other property of the moving ion by evaluating an average function of the property within the channel. For example the average transverse energy is

$$\langle E_\perp \rangle = \int n(E_\perp) f(E_\perp) \, dE_\perp$$

where $f(E_\perp)$ is the variation of $E_\perp$ across the channel and $n(E_\perp)$ is the distribution function.

True steering of ions in channels may not come to equilibrium in distances less than a few thousand Ångstroms and there are obviously a range of scattering events at larger angles than ψ_c which are intermediate between channelling and random collisions. These are referred to as quasi-channelling. Quasi-channelling and the approach to, or from, long range channelling are important because they influence the observed acceptance and emergent angular spread of the ion beam. For example a particle entering in a quasi-channelling direction at the angle ψ_{in} will have a different approach distance to the atomic row given by

$$U(\rho_{qc}) = U(\rho_c) - E\psi_{in}^2.$$

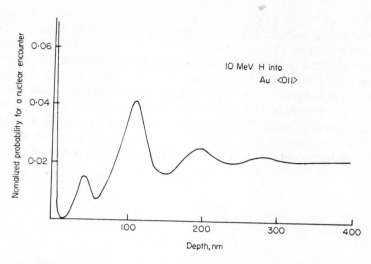

FIG. 3.8. Near surface oscillations in reaction yields from nuclear events for axial channelling (after Barrett, 1971).

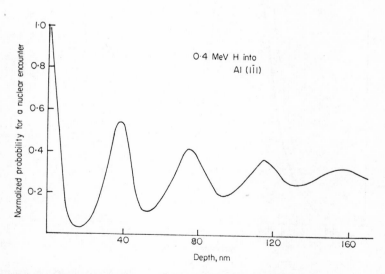

FIG. 3.9. Near surface oscillations in reaction yields from nuclear events for planar channelling (after Barrett, 1971).

Such ions will not be reflected back from the surface, so will be counted as having been channelled, but will differ in their reactions with the lattice compared with truly channelled ions.

Barrett (1971) has calculated the effect of near surface oscillations on the reaction yield from nuclear encounters. Compared with the reaction rate at the surface (i.e. equivalent to a random direction) channelled ions produce fewer nuclear reactions. However, this change in yield shows oscillations and only stabilises after some 2000 Å, as is shown by Figs 3.8 and 3.9. One should note that the stronger potential for axial channelling provides a more dramatic reduction in the yield of events from nuclear encounters.

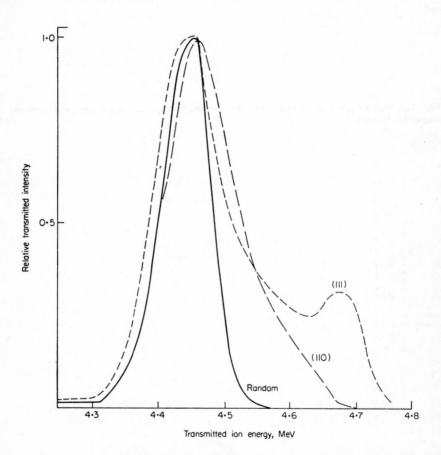

FIG. 3.10. A comparison of transmitted energy spectra for channelled or randomly scattered ions (after Appleton *et al.*, 1967). The primary ions were 4·9 MeV protons incident on a 33μ sample of silicon in a random direction or parallel to (110) or (111) planes.

If channelling persists for an extended distance then one must note a change in the energy distribution of the particles and the distribution function of their location across the channel. The spatial change is to produce a peaking of the flux distribution in the centre of the channel since ions which were not steered into the axial direction will have had more collisons near the critical angle. In real crystals with thermal energy and imperfections this implies that they will have been lost from the channel. "Flux peaking" is experimentally important (Chapter 8) since lattice location studies involving scattering or nuclear reactions require one to know the spatial spread of the beam within the channel.

Similarly the energy distribution of ions within the channel changes with depth. The best aligned ions lose least energy and travel further. This is most obvious if one compares the transmitted ion energy spectra for ions initially incident in channelling and random directions. Figure 3.10 shows results (Appleton *et al.*, 1967) for ions which passed through a silicon specimen. The total number of randomly directed transmitted ions is much less than for the channelling direction but the differences in energy spectra are emphasised by normalising the peak intensities.

In obtaining this energy spectrum Appleton *et al.* (1967) used a wide angle detector to record all the emergent ions. If one uses a small angle detector placed at some angle to the crystal axis then the measured energy spectra may show a series of maxima in the energy spectrum. This is a consequence of the experimental arrangement as is apparent from Fig. 3.11 where we see that not only have we chosen to study channelled ions but in addition we have imposed entrance and exit conditions to the source and small angle detector. This implies a "wavelength" condition on the measured beams. The angle of "bounce" for these beams in turn must define a particular energy, hence an apparent structure in the energy spectrum of the transmitted beam. Such experimental artefacts are not without value and can be

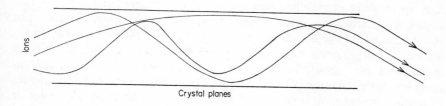

FIG. 3.11. Ion trajectories which emerge at the same angle from the channel may correspond to initially different entrance angles or impact parameters within the channel. A particular length of crystal will select channel "wavelengths" which satisfy the chosen angle of emergence.

used to advantage to study the "end corrections" of the channels as the potential extends beyond the crystal. An example of such data obtained by Datz et al. (1969) is shown in Fig. 3.12.

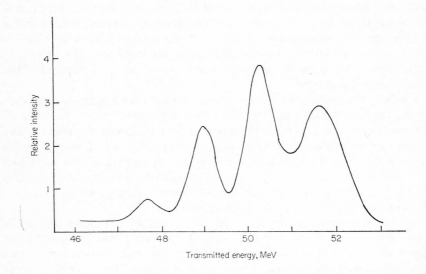

FIG. 3.12. The energy distribution of emitted particles sensed with a crystal of gold tilted 0·5° from the {111} plane. The primary ions were 60 MeV iodine (after Datz et al., 1969).

3.4 THE COMPARISON OF THEORY AND EXPERIMENT

Channelling experiments should give a set of measurements which can be rapidly compared with the theoretical predictions. Indeed it is now relatively easy to measure channelling directions but reliable measurements of half angles, $\psi_{1/2}$, reflection yields or energy spectra are far more difficult. The problem is not in the detection system but in the sample preparation. Problems arise because one requires a very thin single crystal for transmission studies and these should be free of twinning, surface oxides, strain, dislocations impurities or other defects. Whitton (1969) has reviewed the techniques of surface preparation together with dechannelling effects of impurities or radiation damage. He demonstrated the change in channelled range and angular spread of the beam in a series of experiments in which he implanted ions through a thin oxide film. From Barrett's (1971) calculations one expects equilibrium conditions after some 2000 Å into the channel so the presence

3.4 THE COMPARISON OF THEORY AND EXPERIMENT

of monolayers of oxide might be expected to be of minor importance. This is clearly not so as we see from Whitton's (1969) results, Fig. 3.13, of the penetration of krypton ions along the ⟨100⟩ direction in tungsten. The three lines are drawn for an oxide free surface and ones with 100 and 200 Å of oxide. Experimental results which are so sensitive to crystal preparation should therefore be treated with caution even if there is systematic agreement between different experimental groups.

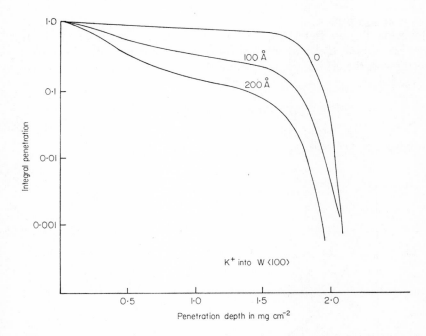

FIG. 3.13. The integral range distributions for ions channelled into tungsten along the ⟨110⟩ axis are influenced by oxide films. The examples show effects from 100 and 200 Å layers (after Whitton, 1968).

It is now necessary to compare the measured angular half width of the channel ($\psi_{1/2}$, see Fig. 3.4) with the critical angle (ψ_c). As a first estimate they should be measuring similar quantities but it is reasonable to introduce a proportionality constant to allow for approximations in the theory, the choice of interatomic potential, thermal effects, and possibly, systematic experimental errors.

The Lindhard (1965) theory is phrased in terms of characteristic angles, and for axial channelling we write

$$\psi_1 = \left(\frac{2Z_1Z_2e^2}{Ed}\right)^{1/2}$$

for high energy ions or

$$\psi_2 = \left[\frac{a\sqrt{3}}{d}\left(\frac{Z_1Z_2e^2}{Ed}\right)^{1/2}\right]^{1/2}$$

for low energy ions.

Ignoring the constants we see that this suggests $\psi_1 \propto E^{-1/2}$ at high energies or $\psi_2 \propto E^{-1/4}$ at low energies. We might now equate ψ_1 or ψ_2 directly with the critical angle ψ_c but at this stage we add correction factors in the form

$$\psi_c = 0.8 \, F_R\left(\frac{1 \cdot 2 u_1}{a}\right)\psi_1$$

where we expect $\psi_c = \psi_{1/2}$.

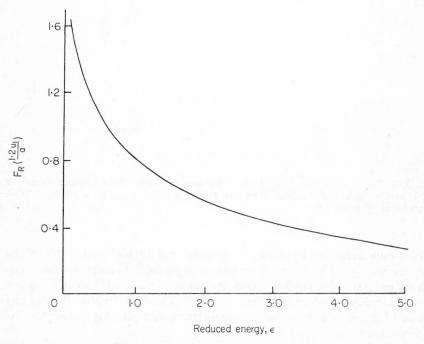

FIG. 3.14. The correction factor $F_R(1 \cdot 2 \, u_1/a)$ for axial channelling.

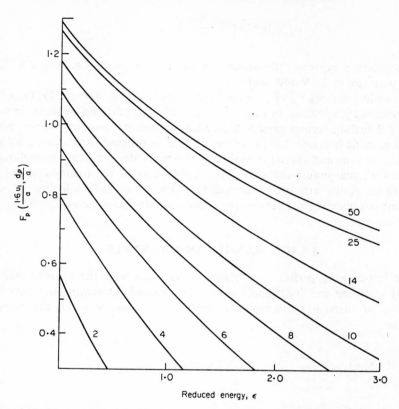

FIG. 3.15. The correction factor F_P ($1 \cdot 6\, u_1/a$, d_p/a) for planar channelling. The set of graphs are for various ratios d_p/a (after Barrett, 1971).

The axial function $F_R(1 \cdot 2 u_1/a)$ includes a thermal correction because u_1 is a measure of the average displacement of the lattice ions and is written in Ångstrom units and a.m.u. as

$$u_1 = 12 \cdot 1 \left[\left(\frac{\Phi(\chi)}{\chi} + \tfrac{1}{4} \right) M_2^{-1} \Theta^{-1} \right]^{1/2}$$

where $\Phi(\chi)$ is the Debye function and Θ the Debye temperature ($\chi = \Theta/T$).

A similar function for planar channelling must account for both thermal vibrations and the relative size of the moving ion and the interplanar spacing, d_p. So in this case

$$\psi_{1/2} = 0 \cdot 72\, F_p \left(\frac{1 \cdot 6 u_1}{a}, \frac{d_p}{a} \right) \psi_a$$

with

$$\psi_a = 0{\cdot}545 \left(\frac{nZ_1Z_2a}{E}\right)^{1/2}.$$

The constant n expresses the atomic density in the plane (i.e. atoms $Å^{-2}$, other units are in MeV and degrees).

The scaling factors F_R of F_p have been computed (e.g. Barrett, 1971) and are presented graphically in Figs 3.14 and 3.15. The addition of these semi-empirical scaling factors produces excellent agreement between theory and experiment, as it must. It is therefore possible to proceed to situations with new sets of ions and crystal directions and predict the expected channelling angles with reasonable certainty. For applications of the theory one may refer to the review article by Gemmell (1974) which includes a listing of the relevant parameters for many of the commonly studied systems.

3.5 ION RANGES IN CHANNELS

If the lattice were perfect then elastic interactions with the channel walls would steer the ion indefinitely along the channel. Dechannelling occurs because of thermal perturbations, impurities or defects on lattice sites,

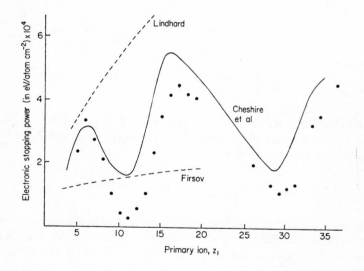

FIG. 3.16. A comparison of the measured electronic stopping power (Eisen, 1968) or Firsov (1958), using Thomas–Fermi atomic models and Cheshire *et al.* (1968) who included electronic shell effects.

interstitials, dislocations or non-elastic processes. Measurements of channelled ranges are therefore subjective with the sample used and the ranges change with implantation dose. The simple theory of channel potentials used a Thomas–Fermi or similar atomic model which replaced the electronic structure by a monoatomic electron density. In channelling events one has minimised the nuclear interactions and energy loss occurs by electronic interactions. There is therefore a possibility of sensing the electronic shell effects which will appear as deviations from the prediction of the Thomas–Fermi potential and will be a function of Z_1. Experimentally such effects were first reported by Eisen (1968) in transmission studies of channelled ions, subsequent theoretical calculations using Hartree–Fock wave functions, rather than a Thomas–Fermi electron distribution, allowed Cheshire *et al.* (1968, 1969) to account for them theoretically. Figure 3.16 indicates the measured and predicted electronic stopping power for a range of Z_1. For comparison the predictions of the simpler theories are also indicated based on the work of Lindhard *et al.* (1963) and Firsov (1958).

In the applications of ion implantation we see that there is no convenient way to predict the range of ions which are initially, or become, channelled. However the fraction of ions which become channelled will be relatively small unless we specifically attempt to enter a channelling direction. It is therefore only necessary to consider if the ion species, target and implant concentrations is likely to be favourable for channelling. If this is not so then the departures in the range profile from the amorphous solid ranges, as estimated in Chapter 2, can be treated empirically.

3.6 CHANNELLING YIELDS

Measurements of the efficiency for back reflection, sputtering, transmission or nuclear reactions all show a variation in yield if the ions become channelled (see Chapter 8). The variation of yield about a channelling direction was sketched in Fig. 3.4 and we saw that adjacent to the channel the yield rose above the normal random direction value. These shoulders represent an enhanced probability of reflection because the ion beam is most likely to have a head-on collision with a surface atom. The minimum yield is of importance since this measures the extent to which the normal continuum potential occupies the channel. The value also changes if there are interstitial ions or dechannelling events. Empirically we may express the minimum yield for a well channelled ion (e.g. Gemmell, 1974) in an axial direction as

$$\chi_{min} = 18 \cdot 8 \, Ndu_1^2 \left[1 + \left(\frac{\psi_{1/2} d}{126 u_1} \right)^2 \right]^{1/2}$$

for low energy ions or

$$\chi_{min} = 18 \cdot 8\, Ndu_1{}^2$$

for high energy ions.

N is the number of atoms per unit volume, other quantities are in Å, MeV or degrees.

REFERENCES

Appleton, B. R., Erginsoy, C. and Gibson, W. M. (1967). *Phys. Rev.* **161**, 330.
Barrett, J. H. (1971). *Phys. Rev.* **B3**, 1527.
Bøgh, E., Davies, J. A. and Nielsen, K. O. (1964). *Phys. Lett.* **12**, 129.
Cheshire, I., Dearnaley, G. and Poate, J. M. (1968). *Phys. Lett.* **27A**, 304.
Cheshire, I., Dearnaley, G. and Poate, J. M. (1969). *Proc. Roy. Soc.* **A311**, 47.
Datz, S., Erginsoy, C., Leibfried, G. and Lutz, H. O. (1967). *Ann Rev. Nucl. Sci.* **17**, 129.
Datz, S., Moak, C. D., Noggle, T. S., Appleton, B. R. and Lutz, H. O. (1969). *Phys. Rev.* **179**, 315.
Eisen, F. H. (1968). *Can. J. Phys.* **46**, 561.
Erginsoy, C. (1965). *Phys. Rev. Lett.* **15**, 360.
Firsov, O. B. (1958). *Sov. Phys. J.E.T.P.* **7**, 308.
Gemmell, D. S. (1974). *Rev. Mod. Phys.* **46**, 129.
Lehmann, C. and Leibfried, G. (1963). *J. Appl. Phys.* **34**, 2821.
Lindhard, J. (1965). *Kgl. Dansk. Vid. Selsk. Mat-Fys. Medd.* **34**, No. 14.
Lindhard, J., Scharff, M. and Schiøtt, H. E. (1963). *Kgl. Dansk. Vid. Selsk. Matt-Fys. Medd.* **33**, No. 14.
Morgan, D. V., (Ed.) (1973). "Channeling", Wiley.
Nelson, R. S. and Thompson, M. W. (1963). *Phil. Mag.* **8**, 1677.
Robinson, M. T. and Oen, O. S. (1963). *Appl. Phys. Lett.* **2**, 30; *Phys. Rev.* **132**, 2385.
Stark, J. (1912). *Phys. Z.* **13**, 973.
Thompson, M. W. (1968). *Contemp. Phys.* **9**, 375.
Tulinov, A. F. (1965). *Usp. Fiz. Nauk.* **87**, 585 (*Sov. Phys. Usp.* **8**, 1966, 864).
Van Vliet, D. (1973). *In* "Channeling" (Morgan, D. V., Ed.), Wiley.
Whitton, J. L. (1968). *Can. J. Phys.* **46**, 581.
Whitton, J. L. (1969). *Proc. Roy. Soc.* **A311**, 63.

CHAPTER 4

Radiation Damage

4.1 INTRODUCTION

Ion implantation adds new atoms to the surface of a material but in addition introduces property changes by disrupting the existing atomic order of the surface layers. This is to be expected since even a light ion will deposit energy at an average rate of 10–100 eV per Å and a heavy ion near the end of its range can lose keV per Å. The atoms of the solid are displaced and whole regions around the track may be treated as though the solid were molten, this is the concept of the thermal spike. If many atoms are displaced then they in turn may displace further atoms, it is customary to refer to this highly disordered region as a collision cascade.

Whilst these dynamic effects are transient, the perturbed structures may re-order into characteristic arrangements of atoms to include vacancies, interstitials and other defects. For the sake of completeness we will list the more common defects, describe their structure and indicate their stability during thermal annealing of the solid in typical situations.

Disorder introduced by ion implantation is called radiation damage or lattice imperfection. One should repeat that the prejudice invoked by the words damage or defects is unfortunate since the changes in structure produced by the passage of fast ions enables us to make materials which were not obtainable by normal thermodynamic processes. It is also true that the "damaged" material may have superior properties to the original solid. Most text books which discuss radiation damage mention defect structures in terms of crystal lattices, because this simplifies both the theoretical and experimental analyses. The major difference between crystals and amorphous solids is that in the former there is long range order of the atoms whereas in amorphous systems the order (i.e. bond length, direction and type of neighbouring atom) exists over a small region. However defect properties also influence a limited region of the solid so their effects are likely to be similar in both crystalline and amorphous solids.

We shall first discuss defects which involve one or two atypical atomic sites and then consider extended defects.

4.2 VACANCIES AND INTERSTITIALS

An unoccupied lattice site is termed a vacancy and an extra atom forced into some region of the lattice is an interstitial. Since these involve relatively small distortions of the original lattice they may exist in "normal" solids and be generated by thermal fluctuations in the lattice. We will first consider the energy required to form an interstitial–vacancy pair in the most favourable case when the lattice can relax during the formation of the defects.

If the solid is ionic or covalent, with directional bonds, then we may consider forming a vacancy by breaking the bonds around an ion and placing it on the surface of the crystal. A surface atom will have about half the normal bond directions satisfied by the lattice so the net expenditure of energy is the energy required to break half the bonds but this is also the sublimation energy of a surface atom. Hence the energy of formation of a vacancy is around 1 eV in an insulator.

In the case of a metal we can make an estimate of the energy for vacancy formation by considering a spherical crystal in which we form a small cavity (the vacancy). Because conduction band electrons spread throughout the solid we can think of the cohesive energy of the metal as being amorphous so the material from the vacancy ($4/3 \pi r_v^3$) can be added uniformly over the outside of the sphere. The total surface area of the metal has increased (by $\approx 4\pi r_v^2$), and we can determine the surface energy, S, of a metal so the energy required to make a vacancy is $U_v \approx 4\pi r_v^2 S$. Numerically this is between 1 and 2 eV for most metals. More sophisticated models predict similar energies of vacancy formation.

We may now calculate the equilibrium vacancy concentration and note that the change in free energy of a material containing vacancies is

$$F = n_v U_v - TS_v$$

where S_v is the entropy change associated with vacancy formation and there are n_v defects. If there are N lattice sites and n_v identical defects the number of different configurations is

$$W = \frac{N!}{n_v!(N-n_v)!}$$

and the entropy of the disorder is $S_v = k \ln W$.

Defects also modify the lattice vibrations. Normally the thermal entropy is

$$S = 3Nk \ln (h\nu/kT),$$

where ν is a vibrational frequency but this will be modified to ν_v near a vacancy.

If each vacancy influences x normal lattice sites the total change in free energy is

$$F = n_v U_v - kT \ln \frac{N!}{n_v!(N-n_v)!} - 3kx \left[\ln \frac{kT}{hv_v} - \ln \frac{kT}{hv}\right].$$

At equilibrium $(\partial F/\partial n_v)_T = 0$, so on substituting the Stirling approximation, $\ln N! \approx N \ln N$, we obtain

$$\frac{n_v}{N} = \left(\frac{v}{v_v}\right)^{3x} \exp\left(-\frac{U_v}{kT}\right).$$

n_v/N is the vacancy concentration. The pre exponential term is not important since v and v_v will differ by only a few per cent and the number of perturbed atoms, x, is generally less than 10. (These statements can be justified experimentally by infra-red spectroscopy, electron spin resonance, field ion microscopy etc.) Thus the Boltzmann factor dominates the defect concentration. The formation energy is important, for in a material at 1000K the concentration would vary from 10^{-5}, if $U_v \sim 1$ eV, to 10^{-50} if the U_v were 10 eV.

A concentration of 10^{-5} defects per normal lattice site is significant as it can represent 10^{18} centres per cm^3. For comparison we may note that impurity atom concentrations are rarely as low as 10^{-5} and even in semiconductor materials a higher purity may only refer to electrically active impurities and not the total impurity concentration. In situations where one can distinguish between the different types of defect or impurity a specific concentration of 10^{-10} may be detectable (e.g. by ESR or luminescence).

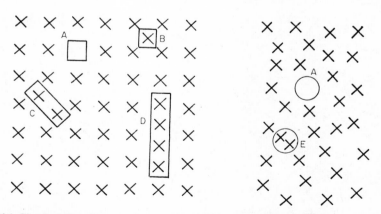

FIG. 4.1. Simple defects in an ordered and an amorphous solid. A is a vacancy, B, C, D are the body centred, split and crowdion models for an interstitial, E is probably interstitial-like in behaviour.

The vacancy concentration introduced in thermodynamic equilibrium at high temperature is not trivial and it is important to note that this concentration may be frozen into the solid by rapid cooling. This situation exists in the formation of many materials. The calculation also indicates the vacancy concentration which could be "frozen" from a thermal spike.

In our estimate of the vacancy concentration we did not distinguish between Frenkel and Schottky vacancy–interstitial pairs. Frenkel defects are those in which both the vacancy and interstitial are included in the bulk of the specimen whereas in Schottky defects the interstitial is considered as being added on to the surface. Figure 4.1 offers some models for vacancy and interstitial sites. The vacancy, A, is an unoccupied lattice site and although some relaxation may occur near the site only short range order is affected, therefore the defect is basically stable and could exist in either a lattice or an amorphous material.

The choice of models for an interstitial is not so obvious nor are the calculations of its stability. Figure 4.1 shows three ways the extra atom may be included in a lattice. Example B is termed a body centred site; C is a split interstitial, that is two atoms sharing a single site and D is a crowdion. Crowdion describes an ion which has been inserted into a normal row of atoms and produces a perturbation over several lattice sites. More realistic sketches of ions in a lattice should include an indication of finite atomic radii. When this is considered, Fig. 4.2, some interstitial models look less promising. In the figure the crosses represent original ion sites. By assuming a hard

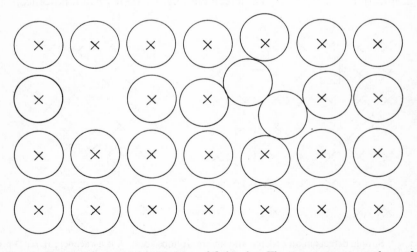

FIG. 4.2. Simple defect models with ions of finite size. The crosses represent the perfect lattice sites.

sphere type ionic potential only the vacancy and the split interstitial could be readily accommodated in this picture.

Even if the defects are formed we should consider if they are thermally stable or if they anneal. The rate of migration will be of the form $v \exp(-U_{motion}/kT)$ where the vibrational frequency, v, is the lattice frequency ($\sim 10^{13} s^{-1}$) and the migration energy will differ for the various defects. In an annealing experiment where the temperature is being steadily increased there will be temperature regions where particular defects anneal.

Table 4.1 indicates the calculated energies of formation and migration for some simple defects in metals together with the temperature at which annealing might occur. As a guideline it is frequently convenient to relate the annealing temperature to the activation energy for migration by $U_m \approx 25 k T_{anneal}$. This estimate is equally useful in insulators or semiconductors where the probability of transition between two electron levels is also significant if $E \sim 25 kT$. Such a guideline is inappropriate if the pre-exponential factor is noticeably different from the lattice vibrational frequency ($10^{13} s^{-1}$). These data were selected from a review of annealing calculations and measurements in the book by Thompson (1969).

TABLE 4.1

Theoretical estimates of the energies of vacancy formation and migration, eV

	Copper	Silver	Gold	Annealing temperature ($U_m/25k$)
Vacancy formation	1·2	1·1	1·0	
Di-vacancy formation	2·2	2·1	1·9	
Vacancy migration	1·0	0·9	—	~ 460K
Di-vacancy migration	0·6	0·5	—	~ 230K

Theoretical estimates of interstitial formation and migration energies in copper, eV.

	Body-centre	split-interstitial	Crowdion
Formation energy	4·0	3·9	4·7
Migration energy	0·05	0·05	0·25
Annealing temperature ($U_m/25k$), K	~ 23	~ 23	~ 115

This data was selected from a review of annealing calculations and measurements in the book by Thompson (1969).

The precise values of these theoretical estimates are open to some debate but they suggest that only vacancy type defects would be generated by thermal processes in these face centred cubic metals. Vacancy or interstitial defects produced during ion implantation will require more energy than the equilibrium calculation since the lattice does not relax during the displacement event. A "standard" estimate of the displacement energy (Section 4.3) is 25 eV rather than the 1 or 4 eV of the equilibrium case. However such quantities of energy are readily available, so all types of defect could be produced. Whether or not they will be stable depends on their migration energy and here the normal lattice calculation is still applicable. The Table 4.1 indicates that vacancy type defects could exist at room temperature but to preserve interstitial defects the damage must be produced and maintained near liquid helium temperature.

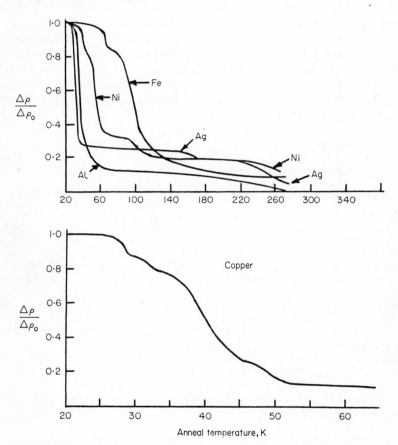

FIG. 4.3. Pulse annealing curves for several metals. The curves record the recovery of radiation induced resistivity after bombardment with 1·5 MeV electrons at 20 K (from Walker, 1962). Measurements were made at 20 K. The structure in the first major annealing stage is shown in more detail for copper.

Typical annealing curves for the annealing of radiation damage in metals is shown in Fig. 4.3 (Walker, 1962). For simplicity the damage was introduced by relativistic electrons (1·5 MeV) which have sufficient momentum to produce point defects but cannot disturb larger regions of the solid. In this example the changes in defect concentration are sensed by electrical resistivity measurements, however, all defects contribute to electron scattering so one cannot distinguish between different species by a resistivity measurement. To separate thermal and defect scattering the resistivity measurements were made at 20 K after successive 10 minute anneals at progressively higher temperatures. The data are presented as the fractional recovery of the resistivity introduced by radiation damage.

This family of curves has the major annealing stage at very low temperatures which is consistent with the theoretical estimates of interstitial migration and the assumption that the bulk of the interstitials anneal by recombining with vacancies. In particularly detailed studies of point defects in copper (e.g. see Corbett and Walker, 1959; Sosin and Neely, 1962 or the review by Walker, 1962) the first annealing stage (16–55 K) has been subdivided into five substages which can be associated with (i) correlated diffusion of interstitials to the adjacent vacancy (ii) random diffusion to other vacancies and (iii) interstitials stabilised by impurities. (Diffusion kinetics will be discussed in Chapter 5.)

Not only do the general features of the annealing curves look similar for all the metals (Fig. 4.3) but also these comments on defect stability are equally applicable to insulators. Once again the interstitial type defects become mobile at much lower temperatures (<50 K) than vacancies (<500 K). One should note that in non-metallic systems the stability of a defect will also depend on the electronic charge state relative to its surroundings. For example in alkali halides the halogen vacancy is stabilised by an electron and can exist up to 500 K whereas without the electron the system is unstable above 50 K.

Whilst it is clear that imperfections exist, together with changes in properties of the solid, it is always a major problem to decide the detailed structure of the defect and its contribution to a particular property. Few measurements discriminate between different defects and one normally resorts to an indirect indentification via a bulk property change (e.g. resistivity, friction, length) by comparison with a theoretical model. Some atomic sites may be located by Rutherford backscattering (see Chapter 8).

In non-metals the directional nature of the ionic or covalent bonds together with the optical transparency of solids which have empty conduction bands combine to make defect identification more positive. The powerful techniques of electron spin resonance (ESR) or electron nuclear double resonance (ENDOR) probe the electron density around a defect and sense the minor shifts in energy levels produced by interactions with nuclear spins. In the classic case of the F-centre in alkali halides (an electron in a halogen vacancy) some 90% of the electron density is seen to be confined to the vacancy. However the technique can still note interactions with neighbours up to six lattice sites from the vacancy. Even if ESR is not applicable the dispersive measurements of optical absorption or luminescence provide some insight into the behaviour of different defect types.

TABLE 4.2 A summary of defects in alkali halides which involve halogen lattice sites. The defects used to be labelled in terms of the optical absorption bands α, F, M, R, etc. but it is now customary to indicate the number of vacancies and the charge state relative to the perfect lattice (i.e. F^+, F, F_2, F_3).

α band (F^+)	A halogen ion vacancy which modifies electronic transitions of the lattice.
F band (F)	A halogen ion vacancy that has trapped an electron. The transitions to higher states are also evident as the bands labelled K, L_1, L_2, L_3.
F_A, F_B bands	(Sometimes labelled A and B)—absorption bands which arise from impurity alkali ions at one of the six neighbouring metal ions sites of an F centre. (There are two bands because of the reduced symmetry of the centre, and the dipole may lie along the axis through, or perpendicular to, the impurity ion.)
F' band (F^-)	A halogen ion vacancy that has trapped two electrons.
M band (F_2)	A pair of adjacent F centres.
R bands (F_3)	A triangular array of three F centres. Although possible arrangements are a linear chain of three vacancies, an L-shaped set in the (100) plane or a triangular set in the (111) plane, the model currently believed is the latter (111) arrangement.
N bands (F_4)	Four F centres linked together. Two possible configurations are a parallelogram of vacancies in the (111) plane which may produce the N_1 absorption and a tetrahedron of vacancies which may give the N_2 absorption.
M', R', N' bands (F_2^-, F_3^-, F_4^-)	By analogy with the F centre the M, R or N centres may trap an additional electron. Some absorption bands have been associated with these defect models.
U band	A hydride ion substituting for a halogen ion on a lattice site. (Hydrogen negative ions may form in interstitial sites (the U_1 band) and hydrogen atoms in interstitial sites (the U_2 band)).

To indicate the range of defects which have been identified, Table 4.2 lists some of the halogen defects in alkali halides. Sketches of the structures are given in Fig. 4.4. It should be realised that alkali halides are particularly simple compounds and yet this table of defects in the halogen sub-lattice

4.2 VACANCIES AND INTERSTITIALS 73

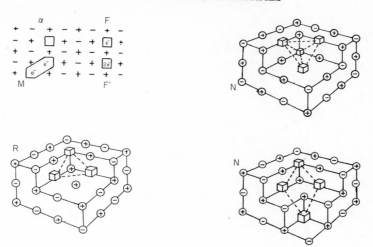

Models of defects in alkali halides

FIG. 4.4. The models of the vacancy defects α, F, F', M, R and N in the alkali halides (see Table 4.2).

represents some 30 years of world wide research. There are many more defects, particularly impurity related centres, for which a "first-guess" model exists. One may ask why no point defects were mentioned for the cation sub-lattice since ion implantation or energetic electrons will certainly produce displacements. It appears that the migration energy is so low that the defects anneal or form into extended defects such as platelets or colloidal metal aggregates. Indeed optical scattering from metal colloids has been observed and one may note that the development of a grain in a photographic emulsion requires the precipitation of a small metallic cluster.

Further examples of the analysis and identification of point defects are given in the general references listed at the end of this chapter. The role of defects in device applications is specifically mentioned in the book by Townsend and Kelly (1973).

To summarise, we are at the stage in point defect studies where we can predict the general behaviour of vacancies and interstitials and in a few materials offer quite detailed justification for the models. However, many defect structures, particularly in metals, can only be surmised from indirect evidence. A realistic view is that we may utilise the property changes of defects introduced by ion implantation without a detailed knowledge of the defect structure. This pragmatic approach is not limited to ion beam technology as the conventional preparation of semiconductors, phosphors, metallic alloys etc. abound with similar uncertainties.

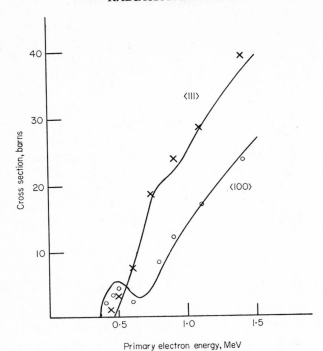

FIG. 4.5. A comparison of theoretical displacement cross sections and resistivity changes for defect formation in iron. Computed values are for $E_D\langle 100\rangle = 20$ eV, $E_D\langle 111\rangle = 30$ eV. (After Lomer et al., 1967).

4.3 DISPLACEMENT ENERGIES

We estimated that to displace an atom from its atomic site would involve an energy of a few electron volts, whereas in a dynamic event the energy required is nearer 25 eV. This displacement energy can be directly measured by noting the rate of defect formation as a function of fast electron bombardment and then comparing this dependence with the theoretical cross section curves, computed for different displacement energies. The cross section for displacement is

$$\sigma = \frac{\pi b^2 (1 - \beta^2)}{4} \left[\frac{E_m}{E_D} - 1 - \beta^2 \ln\left(\frac{E_m}{E_D}\right) \right.$$
$$\left. + \pi \, \alpha\beta \left\{ 2\left(\left[\frac{E_m}{E_D}\right]^{1/2} - 1\right) - \ln\left(\frac{E_m}{E_D}\right) \right\} \right]$$

where

$$b = \frac{2Z_2 e^2}{m_0 v^2}; \quad \beta = \frac{v}{c}; \quad \alpha = \frac{Z_2 e^2}{\hbar c} \approx \frac{Z_2}{137};$$

E_D is the threshold energy for displacement and E_m the energy transferred from the relativistic electron is

$$E_m = \frac{2E_1(E_1 + 2m_0 c^2)}{M_2 c^2}.$$

In an ordered lattice the displacement energy will be sensitive to the direction in which the ejected atom attempts to escape. Such effects have been recorded and in Fig. 4.5 the computed curves for $E_D = 20$ and 30 eV are compared with the resistivity changes produced in thin oriented layers of iron (Lomer et al., 1967) and it is apparent that the theory describes the experimental data. An average E_D of 25 eV seems justified.

For a more extensive treatment of the subject the reader is referred to the book by Corbett (1966). One should realise that electron bombardment is, in terms of property change, just a special case of ion implantation. It has the advantages that defects are produced without any additional ions and the electron range is long compared with the range of an energetic ion, so a larger volume of material may be modified. Range expressions, for unit density material, are given empirically by Glendennin (1948) as

$$R = 407 \, E^{1.38} \quad 0.15 < E < 0.8 \text{ MeV}$$

$$R = 542 \, E - 133 \quad E > 0.8 \text{ MeV}$$

where R is in mg cm^{-2} and E in MeV.

4.4 ALTERNATIVE PROCESSES IN DEFECT PRODUCTION AND SEPARATION

It was assumed, in all the discussions so far, that atomic displacements were the result of elastic collisions and that electronic effects were unimportant. Indeed, elastic collisions are the major reason for defect formation but it should be pointed out that in non-metallic systems defect production rates are influenced by the charge state of the lattice. In particular we must realise that electronic and nuclear losses are not independent events and the state of ionisation of the struck atom and the surrounding lattice atoms may significantly change the energy required for displacement. For example if the ion becomes multiply charged the electrostatic repulsion between it and its neighbours may assist in the removal from the original site. It is also apparent that a multiply charged positive ion will be physically smaller than the singly charged ion and so may require less energy to force its way out from the lattice site. Such an argument implies that the time required for ejection is less than the time needed to recapture electrons and return to the ground electronic state. Defects generated by ionisation have been discussed by Pabst and Palmer (1974) to explain defects induced in silicon during proton and helium irradiation.

We have also assumed that the interstitial–vacancy separation was achieved by a series of displacement events. However a stable separation of the pair can be produced with a smaller expenditure of energy if the lattice will support a focussed replacement sequence (Silsbee, 1957). Figure 4.6(a) describes a row of atoms in which atom B has been struck and received momentum as indicated by the vector. This is sufficient to displace atom C which in turn strikes atom D. The linear progression can be assisted by the ordered nature of the surrounding lattice. Eventually we reach the situation of Fig. 4.6(b) where there is a vacancy at the original site of atom B and the interstitial appears between D and F (in this example as a crowdion).

The process is important because the equilibrium concentration of defects will be higher since correlated recombination of interstitial–vacancy pairs is less likely. In the case of a collision cascade it might also leave the central core rich in vacancies. A situation in which the vacancies might cluster independently of the interstitial processes.

In some materials purely electronic mechanisms can lead to atomic movements. Classic examples are the photographic process in silver halides and colour centre formation in alkali halides. The latter is interesting because it is proposed (Pooley, 1966, Hersh, 1966, Itoh and Saidoh, 1973, Toyozawa, 1974) that the energy is coupled into the halogen sub-lattice by the transient formation of an excited halogen molecule. When this relaxes and then dissociates the energy is directed along a close packed row ⟨110⟩ of halogens

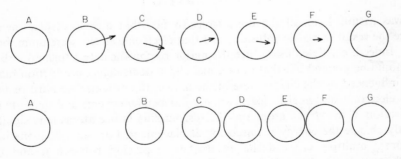

Fig. 4.6. A replacement collision sequence which leads to a vacancy and an interstitial in the row of atoms.

by a replacement collision sequence. To make the process more efficient (i.e. increase the length of the sequence) it was assumed that the collisions occur between small neutral halogen atoms and halogen ions, rather than between pairs of large halogen ions (Smoluchowski et al., 1971). The effective ionic transport is achieved by a parallel process of electronic tunnelling along the collision chain. In alkali-halides the existence of the crowdion interstitial is positively established and there is considerable evidence for this low energy model of defect formation (e.g. Townsend and Kelly, 1973, Townsend, 1973, Saidoh and Townsend, 1976).

In materials which allow these alternative mechanisms of defect production the volume of the solid containing defects may be greater than that expected from the simple elastic collision process.

4.5 DEFECT CONCENTRATIONS

In Chapter 2 we discussed the spatial distribution of the defects produced by ion implantation and realised that it was similar to the concentration profile for implanted ions; and also the ion range and the defect distribution are sensitive to M_2/M_1. Although we discussed the distribution we made no attempt to quantify the number of vacancies and interstitials that were produced, and in the light of the comments in this chapter on the thermal equilibrium of point defects at various temperatures it is not possible to do so without considering a specific example. We could, however, estimate the number of atoms that will be initially disturbed from their original sites, this may not predict the equilibrium concentration of defects, but it is useful (i) in setting an upper limit to the primary defect concentration, (ii) in considering whether or not sufficient disorder is introduced that the solid may undergo a crystalline to amorphous transition, and (iii) the number of displacements near the surface is related to the number of atoms ejected from the surface (i.e. sputtering).

In essence the early models of Kinchin and Pease (1955) Snyder and Neufeld (1955) assumed that no atoms were displaced unless a minimum displacement energy E_D was transferred to the struck atom. If more than $2E_D$ energy were available this would produce a multiplicity of displacements proportional to the energy in excess of $2E_D$. This suggests that a function $v(E_2)$, which measures the number of displacements, has the form of Fig. 4.7. The maximum slope is $E_2/2E_D$, assuming there are no alternative sources of energy loss. The total number of displacements which are made, C_D, is related to the flux, Φ, and irradiation time, t, by

$$C_D = \Phi t \int_{E_D}^{T_{max}} v(E_2) \frac{d\sigma}{dE_2} dE_2,$$

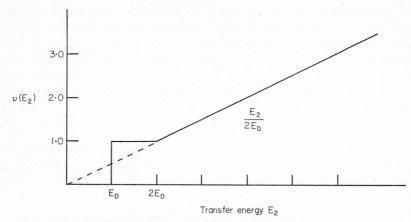

FIG. 4.7. A simple model of the number of displacements $\nu(E_2)$ formed as a function of the transferred energy E_2. E_D is the minimum energy required for displacement.

where $T_{max} = (4M_1 M_2 E_1)/(M_1 + M_2)^2$. This equation can be integrated if we specify $d\sigma/dE_2$. With the usual assumptions about the nature of the repulsive potential for ion pairs (see Chapter 2) the inverse square potential leads to

$$C_D = \Phi t \frac{\pi M_1 Z_1^2 Z_2^2 e^4}{2 M_2 E_1 E_D} \log\left(\frac{4 M_1 M_2 E_1}{(M_1 + M_2)^2 E_D}\right)$$

for energetic light ions. Note that C_D, in this case is approximately proportional to E_1^{-1}. For heavy ions we approach

$$C_D = \Phi t \frac{\pi^2 (0 \cdot 885)^2 M_1 e (Z_1 Z_2)^{7/6}}{4(M_1 + M_2)(Z_1^{2/3} + Z_2^{2/3}) E_D},$$

which is independent of E_1. We would therefore expect a fairly constant density of defects along the track of the implanted ion.

4.6 EXTENDED DEFECTS

Disordered regions, which extend over a large region of a solid, will occur in the form of grain boundaries, dislocations, impurity clusters and crystalline or amorphous inclusions. In compound systems there can also be localized variations is stoichiometry. Such arrays of defects are certainly not passive and, as is explained in undergraduate text books of the solid state, dominate the mechanical properties of solids. Both implanted ions and atomic defects interact with the existing network of extended defects and contribute to their production or removal. Additional large scale features develop in the form of bubbles, if the implanted ions are inert or are gaseous and exceed the solubility conditions existing in the material. The migration of vacancies can generate vacancy "bubbles" called voids, the parallel effect of a solid implant would be termed precipitation. The possible interactions are so diverse and a function of so many parameters that one can only mention typical interactions and leave specific problems for more detailed thought when one knows the defect mobility, alternative phases, sample temperature (see Nelson, 1968).

In the controlled situations of ion implantation these effects can be studied after their production whereas in the similar situation of defects and fission fragments injected by a nuclear reactor it is essential to predict the consequences. The situation is more difficult because in the working life of a reactor one might easily expect every atom to have been displaced as often as 100 times. Ignorance of the consequences, or inability to control them, has led to reactor materials which distorted or failed. Fortunately it has now been realised that reactor conditions can be simulated by ion beam implantation at high dose rates. The radiation damage of the reactor's operating lifetime can then be assessed in days, rather than years.

4.7 FISSION FRAGMENTS AND FISSION TRACKS

The fission of uranium produces energetic neutrons, alpha particles and heavy fission fragments. Initially one considers that the neutrons and alpha particles generate point defects over much of their path and extended defects will occur as a secondary process. For example the inert gases helium ($A = 4$) krypton ($A = 84$) and xenon ($A = 131$) produced by fission may cluster to form gas bubbles (see Section 4.9). Such effects are only important after prolonged irradiations. However the fission fragments are both heavy and enormously energetic. The fission fragment spectrum (Glasstone and Edlund, 1952) peaks near masses 96 and 137 and the ions share up to 200 MeV. The rate of energy dissipation from these highly ionised atoms can be keV per Ångstrom and result in as many as 10^5 target atoms being displaced per fragment. Not surprisingly there is so much damage that a single event leaves an extended defect in the form of a fission track. The disorder and associated strains make the track viewable by chemical etching or electron microscopy. A brief introduction to the subject is given in the book by Chadderton and Torrens (1969).

4.8 DISLOCATIONS

The two basic types of dislocation in crystalline material are the edge dislocation and the screw dislocation. These are shown schematically in Figs 4.8

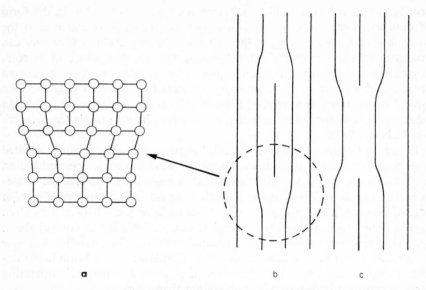

FIG. 4.8. Edge dislocations, which (a), reach the surface, (b) are formed by condensation of a plane of extra atoms and (c) are formed around a collapsed region of the lattice.

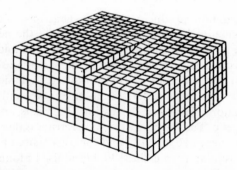

FIG. 4.9. A screw dislocation. The figure shows how a surface step would be generated if part of the crystal is displaced by a lattice unit.

and 4.9. An edge dislocation is the line which separates the parts of a crystal which contain a normal number of atomic planes and a different number of planes (normally ± 1 extra plane). In the sectional view of the solid, Fig. 4.8(a), the dislocation line emerges normal to the plane of the paper. Such a discontinuity can only exist in a lattice structure if the line emerges at the surface or forms a closed path within the solid. (A similar situation of a crack in an amorphous material would not have this restriction.) If many point defects are generated they can form or remove dislocation lines in the following way. For example a large number of interstitial atoms which form into an interplanar "raft", Fig. 4.8(b), may bond as a normal atomic plane in the middle of the raft. However the edge of the plane cannot match the surrounding lattice so an edge dislocation forms. Similarly a raft of vacancies may collapse over the central region which again produces an edge dislocation. The size of the loops can change as interstitials or vacancies migrate in the solid. It is frequently useful to distinguish between the different dislocations and specify the direction and number of extra planes which are involved. This is done by considering a vector path around the region of the dislocation, Fig. 4.10. In a perfect solid the sum of the vectorial steps is zero, but here the resultant is a measure of the dislocation and is called a Burgers vector. For an edge dislocation the vector is perpendicular to the direction of the dislocation.

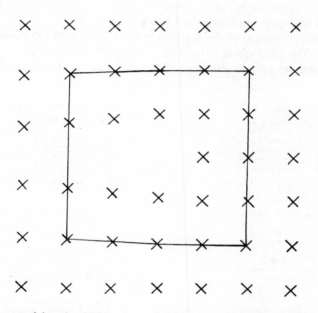

Fig. 4.10. A vectorial path which encloses a dislocation determines the Burgers vector.

In the other basic type of dislocation, the screw dislocation, the Burgers vector is parallel to the dislocation. It can be visualized with the aid of Fig. 4.9, in which part of the crystal has been displaced by an atomic layer. When a screw dislocation intersects the surface there is a surface step and, of course, an atomic site of different chemical stability from the rest of the surface plane.

Such sites are readily resolved in crystals by an electron microscope. In addition they provide sites for whisker growth from the vapour phase and lead to growth spirals around the point of enhanced chemical stability.

One can consider the interaction of dislocations in terms of the vector properties of the Burgers vector. For example a pair of extra planes at one dislocation can be separated to form two separate, or partial, dislocations, each having a halved Burgers vector. Such movement reduces the elastic strain in the lattice so dislocation motion is an energetic possibility. It is this freedom of movement which is so important in mechanical properties of solids. One would expect that shear could take place in a perfect crystal if all the bonds between adjacent planes were disrupted simultaneously. However this would require an enormous energy and the material strength would be some 10^3 greater than is normally observed. This reduction in the strength of normal materials occurs because the dislocation lines allow sideways motion of adjacent layers (Fig. 4.11). Only in dislocation free materials do theoretical strengths approach the observed strengths.

The sideways motion of dislocations is termed slip and the plane in which it moves is the glide plane (Fig. 4.8). As a general rule slip occurs on the plane of densest atomic packing. If the dislocation moves parallel to the

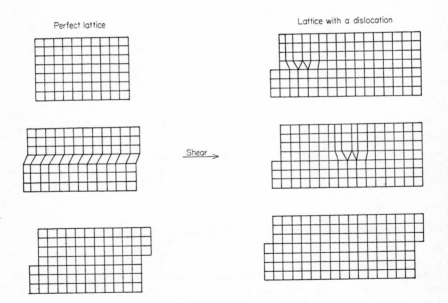

FIG. 4.11. The strength of perfect and faulted material is seen for lattices under shear. The perfect lattice requires simultaneous breaking of all bonds between adjacent layers whereas the dislocation region allows the shearing faces to cross one atomic plane at a time.

extra atomic plane it is called climb. This is only possible if the dislocation is a source or a sink for point defects, such as those generated during ion implantation. Studies of dislocations made in an electron microscope during ion bombardment (Nelson and Mazey, 1973) reveal rapid activity in the dislocation network as existing lines sweep through the sample and new dislocations are generated.

Not all parts of the dislocation line are equally mobile and the line may be pinned at nodes where dislocation lines intersect. These points may then act as sources of vacancies or interstitials if the line is forced to bow through the crystal whilst anchored at fixed points. Pinning may also occur by the presence of impurity atoms. The strain field around a dislocation together with the altered chemical stability make dislocation lines favoured sites for the accretion of lattice impurities.

If a segment of a dislocation line is strongly pinned then applied stresses may make it the source of dislocation loops. Such a mechanism suggested by Frank and Read (1950) is illustrated by Fig. 4.12. Here the pinned segment bows under the stress across the slip plane and expands by the inclusion of more material. Once the line reaches a semicircular shape it will expand beyond the pinning points so as to enclose them. However if the sections

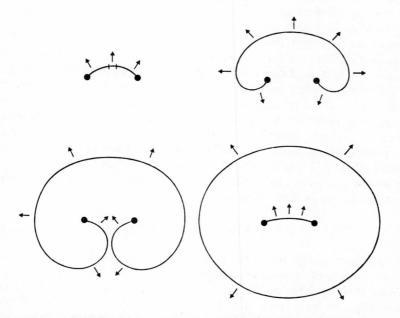

FIG. 4.12. A mechanism proposed by Frank and Read (1950) for a strongly pinned region of a dislocation line to act as a source for dislocation loops.

from each side of the loop intersect they will annihilate since they have opposite Burgers vectors. This will leave a complete loop and the original pinned segment. One can therefore generate a series of loops based on the central segment and examples are frequently observed. In crystalline materials the strain energy modifies the circular loops into a series of lines which follow low energy crystallographic directions. Loops formed from a Frank–Read source will be inhibited once the outer loops pile up at a grain boundary.

It is clear that in repeated stress cycles dislocations will be generated which can form a network and so "work harden" the material. Ion implantation and the associated radiation damage can play a role in controlling the rate of hardening in a surface layer. Generally one would expect the dislocation network to be more heavily pinned in the implanted region and to provide a toughened surface.

If a high concentration of ions are injected into the solid it may be appropriate to think of the system as an alloy. Whilst a discussion of metallic alloys is outside the scope of this book the general features of alloy hardness may be appreciated. If the additional ions are randomly distributed then the strength of the material is similar to the pure system. This situation can be achieved during implantation without perturbing the original phase of the solid whereas random impurity dispersion by thermodynamic means (i.e. solution) may require a quenching procedure from a high temperature phase.

If the alloy is maintained at a temperature where the impurities can migrate then they may cluster, move to dislocations, or form new alloy phases. Such effects are referred to as ageing. In the case of precipitation hardening the first stage may be the formation of small precipates which pin the dislocation network, and so increase the hardness. However prolonged ageing can cause the precipitate regions to grow and then the dislocation lines can effectively pass between these "islands" and the material is again reduced in strength. Annealing of alloy systems must be carried out with these possibilities in mind. The main advantage of ion implantation over chemical alloying is that the solubility phase diagram can be bypassed to a large extent. It also allows one to harden just the surface layers of a solid without a high temperature heat treatment.

4.9 BUBBLES AND VOIDS

The migration energy for clusters of vacancies is higher than for point defects so if a high vacancy concentration is produced in a solid there is a tendency to form large empty volumes, or voids. Similarly, insoluble gas atoms diffuse to these voids. If there is gas contained within these volumes then they are referred to as bubbles. The shape of the bubble is determined by the surface energy of the walls so in a crystalline structure they are polyhedral. Because bubble formation requires nucleation sites the bubble density will vary with the impurity content of the solid. It is also obvious that the competitive process of vacancy capture at nucleation sites, or annihilation by interstitial recombination, etcetera, will produce different equilibria under different conditions of temperature and irradiation flux. Hence bubble formation will also depend on the rate of defect production.

A bubble contains many more vacancies than gas atoms because the walls are stabilised against collapse by the gas pressure. Barnes (1963) suggested that if the bubble had an internal pressure, p, volume $v = 4/3\,\pi r^3$, a surface area $a = 4\pi r^2$ and a surface energy γ per unit area, then in equilibrium the changes in surface energy equals the work done in changing the volume,

i.e. $\quad p\,dv = \gamma\,da$

or $\quad p4\pi r^2 = \gamma\,8\pi r\,dr$

so $\quad p = \dfrac{2\gamma}{r}.$

If the bubble behaves as a perfect gas $pv = 3/2\,mk\,T$ so

$$m = \left(\frac{8}{3}\frac{\pi\gamma}{kT}\right)r^2.$$

Barnes and Mazey (1973) justified this relationship by noting that when n bubbles coalesced the initial and final radii satisfied the equation

$$\Sigma_n r_n^2 = R_{\text{final}}^2.$$

As a result of either ion implantation or nuclear fission the total number of ions in the solid is increased. There is therefore an expansion of the solid with the fractional change in volume being proportional to the mean bubble radius (e.g. Thompson, 1969). To minimise the amount of swelling it is essential to have the gas contained in a large number of small bubbles. This is possible if the initial material contains suitable impurities which can act as nucleation sites.

Bubbles may diffuse through the material and coalesce so the problem of swelling is increased at higher temperatures where this movement occurs. At very high temperatures the problem is reduced, in part because more of the irradiation damage anneals before bubbles form (Mazey et al., 1971).

If bubbles diffuse to the surface then the internal gas pressure can cause them to burst and fragments of the surface are removed. This is termed blistering. Such problems may not seem relevant to normal ion implantation work except in the simulation of reactor damage, however, blistering does occur under certain circumstances even if small quantities of implanted material are involved. The problem can arise when a masking or encapsulating layer has been deposited on the surface. If the specimen is annealed after implantation the trapped gas, or gas evolved from decomposition of the substrate, will attempt to diffuse from the specimen. If the encapsulating layer stops the diffusion of the gas, or the layer to substrate adhesion is poor, then the gas will collect at this interface. The resultant bubbles and surface blistering can then destroy the encapsulating layer.

4.10 SUMMARY

In this chapter we have made a rapid survey of typical point and extended defects which are likely to be encountered in the course of ion implantation. Whilst it was possible to indicate a few general consequences of the defect interactions one must specify many parameters before making predictions for a particular system. The literature of radiation damage, dislocations, metallic alloys etcetera is very extensive and a few general references are included with this chapter.

REFERENCES

Barnes, R. S. (1963). UKAEA Rept. AERE R 4429.
Barnes, R. S. and Mazey, D. J. (1963). *Proc. Roy. Soc.* **A275**, 47.
Chadderton, L. T. and Torrens, I. M. (1969). "Fission Damage in Crystals" Methuen.
Corbett, J. W. (1966). Supp. 7, *Solid State Physics*.
Corbett, J. W. and Walker, R. M. (1959). *Phys. Rev.* **115**, 67.
Frank, F. C. and Read, W. T. (1950). *Phys. Rev.* **79**, 723.
Glasstone, S. and Edlund, M. C. (1952). "Nuclear Reactor Theory", Macmillan.
Glendennin, L. E. (1948). *Nucleonics*, **2**, 12.
Hersh, H. N. (1966). *Phys. Rev.* **148**, 928.
Itoh, N. and Saidoh, M. (1973). *J. de Phys.* **34**, C9, 101.
Kinchin, G. H. and Pease, R. S. (1955). *Rep. Prog. Phys.* **18**, 1.
Lomer, J. N. and Pepper, M. (1967). *Phil. Mag.* **16**, 1119.
Mazey, D. J., Hudson, J. A. and Nelson, R. S. (1971). *J. Nucl. Mat.* **41**, 257.
Nelson, R. S. (1968). "The Observation of Atomic Collisions in Crystalline Solids", North Holland.
Nelson, R. S. and Mazey, D. (1973). "Ion Surface Interaction, Sputtering and Related Phenomena", p. 199 (Behrisch *et al.,* Eds), Gordon and Breach.
Pabst, H. J. and Palmer, D. W. (1974). *Rad. Eff.* **21**, 135.
Pooley, D. (1966). *Proc. Phys. Soc.* **87**, 245.
Saidoh, M. and Townsend, P. D. (1976). *Rad. Eff.* (To be published.)
Silsbee, R. H. (1957). *J. Appl. Phys.* **28**, 1246.
Smoluchowski, R., Lazareth, O. W., Hatcher, R. D. and Dienes, G. J. (1971). *Phys. Rev. Letts.* **27**, 1288.

REFERENCES

Snyder, W. S. and Neufeld, J. S. (1955). *Phys. Rev.* **97**, 1636, *ibid* **99**, 1326.
Sosin, A. and Neely, H. H. (1962). *Phys. Rev.* **127**, 1465.
Thompson, M. W. (1969). "Defects and Radiation Damage in Metals", Cambridge University Press.
Townsend, P. D. (1973). *J. Phys. C.* **6**, 961.
Townsend, P. D. and Kelly, J. C. (1973). "Colour Centres and Imperfections in Insulators and Semiconductors", Sussex University Press.
Toyozawa, Y. (1974). Proc. of Int. Conf. on Colour Centres, Sendai.
Walker, R. M. (1962). "Radiation Damage in Solids", Academic Press.

CHAPTER 5

Diffusion and Thermal Annealing

5.1 DIFFUSION AND ANNEALING

Ion implantation can overcome chemical barriers and solubility rules and place atoms in a target material which would not be allowed by other processes. The energetic atom however eventually comes to rest and is then subject to the lattice forces and may move a considerable distance from its initial resting place, particularly if the specimen is annealed. Such migration of an atom through a solid by random motion is called diffusion and is of course not specific to ion implanted systems. The diffusion equations are transport equations and were initially derived from Fourier's heat conduction equations by Fick who assumed that the quantity, Q, of material diffusing across unit cross section was proportional to the concentration gradient measured normal to the section

$$Q = -D\frac{\partial C}{\partial x}. \tag{5.1}$$

The proportionality constant, D, is called the diffusion coefficient and has the dimensions $cm^2\ s^{-1}$. This equation is called Fick's first law. If we consider a volume of the solid bounded by two such parallel unit surfaces a distance dx apart, the quantity diffusing in through one surface is given by equation (5.1). The out diffusion through the second surface is

$$Q + \frac{\partial Q}{\partial x} = -D\frac{\partial C}{\partial x} - \frac{\partial}{\partial x}\left(D\frac{\partial C}{\partial x}\right)$$

whence

$$\frac{\partial Q}{\partial x} = -\frac{\partial}{\partial x}\left(D\frac{\partial C}{\partial x}\right).$$

But $\partial Q/\partial x$ equals the negative rate of change of concentration, $-\partial C/\partial t$, hence we have Fick's second law

$$\frac{\partial C}{\partial t} = \frac{\partial}{\partial x}\left(D\frac{\partial C}{\partial x}\right), \tag{5.2}$$

which is linear in the time derivative and hence irreversible which makes for difficulties in the thermodynamic treatment of the problem, Le Claire (1949). Under the usual assumption that D is independent of concentration (5.2) becomes

$$\frac{\partial C}{\partial t} = D \frac{\partial^2 C}{\partial x^2}. \qquad (5.3)$$

For the ideal dilute solution case Einstein has shown that

$$D = \mu kT,$$

where μ is the mobility of the diffusing species, that is, the average velocity under unit force.

Each diffusing species has its own diffusion coefficient and in a non-isotropic medium the diffusion coefficients will be direction dependent.

Because of its importance in all interface and mixed system problems, diffusion has been extensively studied for decades, and the work has been collected in a number of books, for example, van Bueren (1961), Gertsniken and Dekhtyar (1960), Boltaks (1963). As atoms on surfaces or near extended defects, such as grain boundaries, dislocations or voids, are less constrained than atoms in the perfect solid they will be more mobile and hence display enhanced diffusion. The same considerations apply to an ion implanted region. Where the lattice is most disordered the rate of diffusion will be greatest and may differ considerably from the value obtained by the conventional diffusion technique of evaporating a layer of the diffusing material onto a surface of the substrate and heating for a sufficient time to achieve measurable penetration. Moreover, the rate of diffusion may differ with the amount of damage and thus an implant profile will change on annealing the specimen.

The techniques of anodic stripping and polishing off thin layers, usually in conjunction with radioactive tracers, to determine diffusion or implant profiles has been described in Mayer et al. (1970) and numerous references to specific systems will be found there. We will confine ourselves to backscattering measurements and discuss effects which are of particular importance over the small distances encountered in thin films and in ion implanted layers. The sensitivity of this technique is also useful in extending diffusion measurements to lower temperatures where the penetration distance may be too small to be resolved by other methods. Backscattering, unlike sectioning methods, is nondestructive, allowing annealing to be repeated and re-measured. There may however be beam interaction effects, particularly with charged defects which may be present in insulators and semiconductors, (Bourgoin and Corbett, 1972). (For those readers who are unfamiliar with backscattering measurements it may be appropriate to digress to Chapter 8.)

The change in diffusion with damage is not always a deleterious effect, it has been used to rapidly move unwanted impurity atoms away from electrically active regions, a process called damage gettering, and to modify implant profiles for the better in a number of systems.

5.2 BACKSCATTERING MEASUREMENTS OF DIFFUSION

The general solution of equation (5.3) for the case

$$\lim_{x \to 0} \left(\frac{dC(x,t)}{dx} \right) = 0 \tag{5.4}$$

is given by Boltaks (1963),

$$C(x,t) = (4\pi Dt)^{-1/2} \int_0^\infty C_0(y) \, [\exp - (y-x)^2/4Dt \\ + \exp - (y+x)^2/4Dt] \, dy \tag{5.5}$$

where $C_0(x) = C(x, 0)$ is the profile at $t = 0$. Myers et al. (1974) have used this expression to obtain diffusion coefficients for Cu implanted at 100 keV in Be by approximating the initial profile by the Gaussian

$$C_0(x) = K(\pi Dt_0)^{-1/2} \exp - x^2/4Dt_0 \tag{5.6}$$

with K and t_0 adjustable constants. From (5.5) and (5.6),

$$C(x,t) = K[\pi D(t + t_0)]^{-1/2} \exp - x^2/4D(t + t_0) \tag{5.7}$$

or

$$[W(t)]^2 = 4Dt \ln 2 + [W(0)]^2 \tag{5.8}$$

where the profile width $W(t)$ is defined by

$$C(W, t) = \tfrac{1}{2} C(0, t)$$

The diffusion coefficient D was obtained from a $[W(t)]^2 : t$ plot. Such a plot for their measurements for a 420°C anneal are given in Fig. 5.1. It shows the profile width rapidly broadening to 2000 Å, pausing and then obeying equation (5.8) with a slope that gives the diffusion coefficient of $D = 8\cdot 5 \times 10^{-16}$ cm^2 s^{-1}. The initial anomalous diffusion is almost certainly associated with damage produced predominantly near the end of the implanted ions trajectory. For their backscattering measurements Myers et al. used 2 MeV He$^+$ which give a depth resolution of about 0·03 µm down to 2 µm and

2 MeV protons which resolve about 0·3 μm down to a depth of about 10 μm in Be. This is sufficiently good resolution to reveal diffusant trapping effects that may lead to errors in determining D, for example the formation of oxides, intermetallic compounds or precipitation when the local solubility is exceeded. They found that oxidation leads to a BeO layer from which Cu was excluded.

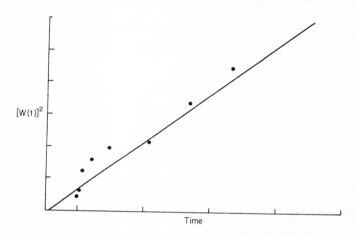

FIG. 5.1. A plot of the square of the width of the Cu peak profile against time, for Cu implanted at 100 keV into Be, annealed at 420°C. Note the anomalous initial diffusion which corresponds to diffusion over about the first two thousand angstrom. The straight line gives a diffusion coefficient for Cu in Be of $8·5 \times 10^{-16} \text{ cm}^2 \text{ s}^{-1}$, determined largely by points of longer t which are not shown here.

Oxidation may be seen from a backscattered spectrum and both the amount of oxygen and its location can be calculated (see Chapter 8). In Fig. 5.2, (Myers et al., 1974) obtained with 2 MeV He$^+$ backscattered at 170°, the diffusion of the initial localised implanted Cu away from the surface of the Be is clearly seen, as is the increase in oxygen from 6×10^{16} atoms cm^{-2} to 2×10^{17} atoms cm^{-2} after the 36 minute anneal at 595°C.

We should note that the Cu–Be system is a particularly favourable one for backscattering. The energy of the particle backscattered by 100° is $E_1(M_2 M_1)^2/(M_2 + M_1)^2$. For Be, $M_2 = 9$ and the backscattered energy of the He is low. For Cu, $M_2 = 64$ and the backscattered He has much more energy leaving a considerable gap along the energy axis, Fig. 5.2, in which peaks due to intermediate masses such as C and O can be clearly measured. The Cu can also diffuse a considerable distance into the Be before the tail of the diffused profile is buried in the edge of the Be spectrum.

The less favourable case of the interdiffusion of Cu, $M_2 = 64$ and Au, $M_2 = 197$ has been recently studied by Borders (1973). Fig. 5.3 shows his

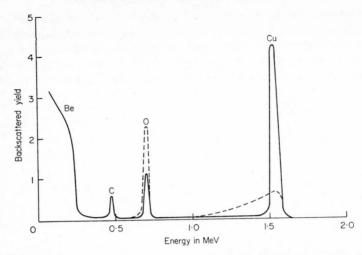

FIG. 5.2. Backscattered spectrum for Cu implanted in Be at 100 keV. 2 MeV He$^+$ backscattered at 170°. The intermediate mass O and C peaks are clearly resolved between the Cu peak and the Be edge. The dashed curve shows the effect of annealing at 595°C for 36 minutes as both an increase in oxidation and a spreading of the Cu peak.

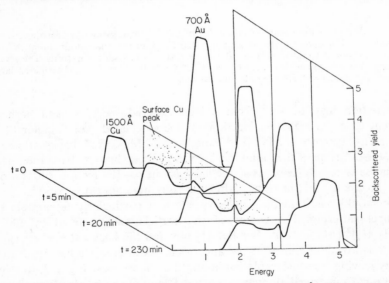

FIG. 5.3. Backscattered spectrum for a 700 Å layer of Au on 1500 Å of Cu on a sapphire substrate. The substrate edge is not shown. No interpenetration is apparent in the Au evaporated spectrum. The Cu and the Au rapidly interpenetrate when annealed at 225°C. The small peak that appears, even after a 5 min anneal, corresponds in energy to Cu on the outer surface of the Au and is the result of extremely fast grain boundary diffusion through the overlaying Au layer. Both energy and yield scales are in arbitrary units.

backscattered spectrum from a 700 Å Au layer on 1500 Å of Cu on sapphire. The as-evaporated spectrum shows an absence of interpenetration before heating. Diffusion at 225°C rapidly produces overlap of the Cu and Au peaks.

For these thin films grain boundary diffusion was found to be dominant in the 200–500°C region with the diffusing species moving rapidly down between the grains and then more slowly penetrating the grains, probably forming intermetallic compounds, such as Cu_3Au or $CuAu_3$, at the grain edges. Enhanced diffusion in the Cu–Au system was in fact found in the pioneering application of backscattering to diffusion measurements by Sipple (1959). He used 90° scattering protons and deuterons and a magnetic spectrograph which had the excellent resolution of 4 keV. However, this instrument recorded photographically and lacked the speed, convenience and ready availability of the modern solid state detectors.

5.3 DIFFUSION IN THIN FILMS

One of the reasons why it has taken so long since Sipples's measurements for backscattering to be generally extended to many metallic systems has been the concentration, by laboratories that might have done the work, on semiconductors, and the lack of a pressing need for information on diffusion over the small distances of which the method is capable. It is moreover easier to get reproduceable results on single crystals of high purity silicon than on notoriously variable thin metal films.

With the need to make solid state devices smaller and smaller the technological spur is now being applied and more laboratories are turning their attention to diffusion and annealing behaviour over distances of a few hundred angstroms and interesting problems are arising. For example, gold is commonly used as a conductor in integrated circuits as it is a good conductor and does not easily corrode. However its inertness reduces its adhesion compared with metals which form adherent oxides and can therefore chemic-

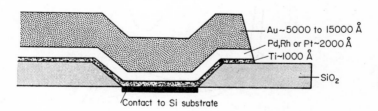

FIG. 5.4. Multilayer evaporated conductor typical of the kind used on semiconductor devices. The Ti forms a strong bond to the substrate; the intermediate Pd, Rh or P minimises intermetallic corrosion between the Ti and the Au and the top Au layer has a high conductivity and minimum corrosion.

ally bond to SiO_2. Ti between the Au and the SiO_2 increases the adhesion but unfortunately the corrosion between the Ti and the Au also greatly increases and so a barrier of Pd, Rh or Pt is added between the Ti and the Au, (English and Turner, 1972). We now have the complex layer system shown in Fig. 5.4 and it is clear that interdiffusion over distances of a few hundred angstrom may seriously effect the properties of the system.

The vacuum annealing behaviour of the Ti:Pd:Au and Ti:Rh:Au systems have been studied by DeBonte et al. (1973) using Rutherford backscattering. One of the complications which occurs with such a three layer sandwich is that diffusion into the central layer from one side may block the diffusion from the other side. It was found, for example, that in the Ti:Pd:Au system the Au rapidly diffused into the central Pd layer, blocking the diffusion of Ti. The interpretation of the data is complicated by the fact that during annealing, grain boundary diffusion, bulk diffusion and the grain size all change differently as a function of temperature. One might also expect that the rate of grain growth would be influenced by foreign atoms diffused into the grain boundaries but little relevant information seems to be yet available. DeBonte et al. (1973) have looked at their films in the transmission electron microscope and believe that this effect was not a major one for their systems.

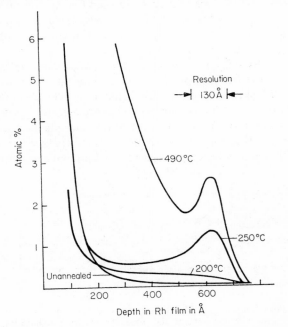

FIG. 5.5. The diffusion profile of Au into Rh in the substrate: Ti: Au: Rh system. The Au: Rh interface is at 0 on the depth axis and the outer surface of the Rh at 650 Å.

Depth profiles for Au diffused into Rh in the Ti:Au:Rh system are shown in Fig. 5.5. The 200°C profile shows Au diffusing rapidly through the Rh grain boundaries. The onset of significant bulk diffusion produces the profile at 250°C, where the peak at the Rh surface, corresponding to $1 \cdot 3 \times 10^{15}$ Au atoms cm^{-2}, about a monolayer, is thought to be due to the Au which diffused rapidly through the Rh grain boundaries and spread over the Rh surface. The 490°C curve shows dominance by bulk diffusion. Making reasonable assumptions about grain boundary size and distributions they deduce the bulk diffusion coefficient of Au into Rh as 5×10^{-7} cm^2 s^{-1} with an activation energy of $1 \cdot 34$ eV and for Rh into Au 9×10^{-9} cm^2 s^{-1} and 1 eV.

Diffusion analyses by backscattering are not limited to metallic systems for example Myers (1974) studied the interdiffusion of hydrided lithium layers in a $LiOH-Li_2O-LiH$ sandwich.

5.4 STRESS ASSISTED DIFFUSION

For implanted regions we have lattice damage to consider in addition to the effects mentioned in the above section. However, even for the interdiffusion of evaporated layers stresses are expected to exist at the boundaries between the intermetallic compounds and the base metal (Dearnaley and Hartley, 1974). The relaxation of these stresses may produce cracks, voids and dislocations which will have a large influence on diffusion, particularly at low temperatures. If the intermetallic compound is hard, brittle fracture may result, a fact well-known in corrosion which is discussed in Chapter 10.

Eer Nisse (1974) has measured the stress produced in a number of insulators and metals implanted with protons and He^+. He used a cantilever beam technique where a long thin sample is irradiated on one side and the resultant beam bending measured by a capacitance change method, (Eer Nisse, 1971) and a quartz resonator method (Eer Nisse, 1972) with the implanted film deposited on a quartz crystal whose resonant frequency is shifted by the stresses produced on implanting. For bombardment with 120 keV He^+, insignificant stress build up was found for Cu and more detected for Au; Erbium films showed a steady increase in stress with dose, Fig. 5.6, up to a fluence of 10^{16} cm^{-2}. The peak stress is approximately the integrated lateral stress divided by the range (~ 6000 Å), which comes out to be about 10^9 dyne cm^{-2}. The stress relief above 10^{16} cm^{-2} may be elastic plastic yield in the Er but could equally be flow in the Au electrode layer which underlies the Er.

Evidence for very large stresses has also been found by a cantilever beam method for nitrogen or argon implanted to doses of 10^{17} or 10^{18} ions cm^{-2} in nitriding steel (Hartley, 1975).

Similar results occur for protons in fused silica with some evidence for saturation at 2×10^5 dyne cm^{-1} integrated stress at a fluence of 10^{16} cm^{-2}, except that in this case irradiation causes contraction rather than expansion, as was to be expected from photo-elastic measurements, (Primak, 1969). The more interesting results come from Ag–phosphate dosimetry glass implanted with protons, He$^+$ and O$^+$ where a good fit is obtained between experiment and calculations based on the work of Brice (1971), assuming independence of damage from ionizing and from direct displacement processes. This offers the hope of relating the relative energy deposited into the two processes.

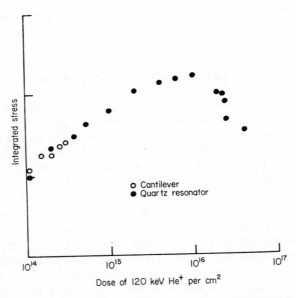

FIG. 5.6. Integrated stress as a function of ion dose, for Er films irradiated at 120 keV with He$^+$. At low dose there is good agreement between the cantilever beam and quartz resonator methods. Stress recovery occurs above a dose of 10^{16} ions cm^{-2}, (Eer Nisse, 1974).

5.5 THE STOICHIOMETRY OF THIN FILMS

Oxidation of films frequently occurs during annealing or even during evaporation or implantation. This can have a profound effect on subsequent diffusion and we should therefore measure the amount, location and stoichiometry of any oxide layers. In thin silicon oxide films memory and switching effects have also been detected which are related to the stoichiometry of the films. Resonance reactions have been used extensively, for example Amsel (1973)

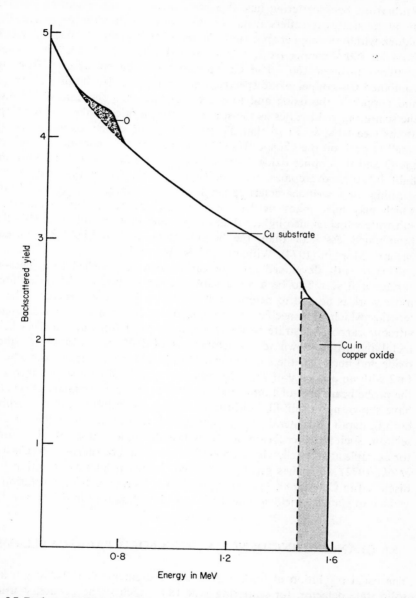

Fig. 5.7. Backscattered spectrum from copper oxide on Cu using 2 MeV He$^+$ backscattered at 160°. From the distribution of the O and the Cu in the oxide the stoichiometry of the copper oxide can be deduced.

studied oxidation and this technique will be discussed in Chapter 8. Rutherford backscattering has also been used but, for oxidation of thick metal substrates it suffers from the disadvantage that the oxygen peak is hidden within the copper spectrum. Figure 5.7, (Morgan, 1974) shows results from an 800 Å copper oxide film measured with 2 MeV He$^+$ ions backscattered through 160°. The Cu spectrum can be substracted from the combined Cu–copper oxide spectrum and hence the distribution of oxygen and copper in the oxide and thus the stoichiometry determined. However the scattering yield varies as the square of the atomic numbers of the target atoms (see Chapter 8) so that $Y(Cu)/Y(O) = 13.2$ which accounts for the small O peak on the Cu spectrum. The errors are thus considerable in both the O and the copper oxide distribution. The use of a thin metal film on a light substrate overcomes this disadvantage (Fig. 5.2, at the cost of substituting for a well-characterised single crystal surface, an evaporated layer which may have many of the uncertainties we have discussed above). A substrate which is also lighter than oxygen and more readily available than (and much less toxic than) the Be substrates used in Fig. 5.2 is vitreous carbon (Morgan (1974), Williams and Kelly (1975)).

This recently developed form of carbon can be polished to a glass-like surface and seems to be a good substrate for evaporated layers although more work is needed to determine if the layers have any structural characteristics which are specific to this substrate. Morgan (1974) has used a vitreous carbon substrate to analyse a number of films and detects heavy metal impurities down to concentrations of 10^{18} atoms cm^{-3} in silicon oxide and finds that the stoichiometry of his SiO$_x$ films varies from $x = 0.9$ to 2 with an accuracy of 3%. The measurements is of course averaged over the probe beam area of 1 mm^2 and in view of the grain boundary effects that have shown up in thin film diffusion a higher resolution would be desirable both in depth and laterally by using a backscattering beam of smaller cross section. Such ion microbeams are now available in a number of laboratories, for example at Harwell, where a 5 μm ion beam has been developed (Cookson et al., 1972) which has greatly improved the usual lateral resolution (also discussed in Chapter 8). The other problem of increased depth resolution has yielded to glancing incidence methods as will be described in the next section.

5.6 GLANCING INCIDENCE RUTHERFORD BACKSCATTERING

The usual resolution of backscattering measurements is 150–200 Å using a solid state detector, for scattering near 180° which is the geometry usually employed. However, as Bøgh (1968) showed, this is not the best geometry for maximum depth resolution. For near surface backscattering with a

5.6 GLANCING INCIDENCE RUTHERFORD BACKSCATTERING

system energy resolution of ΔE, the depth resolution dz is given by

$$dz = \Delta E/(k^2 S_1/\cos\phi_1 + S_2/\cos\phi_2)$$

where k is a kinematical factor depending on the mass of projectile and target atoms, ϕ_1 and ϕ_2 the angles between the surface normal and the entering and measured backscattering beams repectively and S_1 and S_2 the

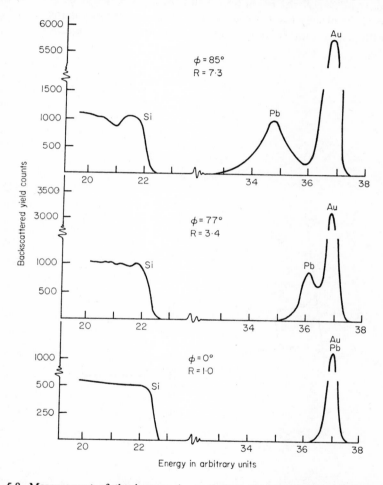

FIG. 5.8. Measurement of the increase in resolution obtained by backscattering at an increasingly oblique angle. The bottom results are for incidence and backscattering at near normal incidence ($\phi = 0°$). The middle spectrum is for incidence and backscattering at 77° to the normal, the resolution is increased 3·4 times and the Pb peak has emerged. An increase in the angle to 85° increases the resolution at normal incidence and clearly resolves the Pb and Au peaks.

stopping powers in and out which are both a function of the projectile energy. By making both ϕ_1 and ϕ_2 75–85° instead of nearly zero in the usual normal incidence experiment the resolution of the technique is increased to about 20 Å, almost an order of magnitude improvement (Williams, 1974). The increased resolution is clearly shown in Fig. 5.8 where near normal incidence is unable to resolve Pb implanted in Au on a Si substrate. At $\phi = 77°$ the resolution is improved by a factor of 3·4 and the Pb shoulder is clearly resolved; at $\phi = 85°$, the resolution enhanced by 7·3 and the Pb peak is almost completely resolved from the Au peak.

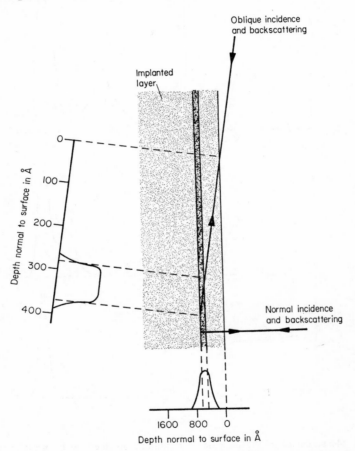

FIG. 5.9. For thin layers probed at normal incidence the energy (and hence depth) resolution of the detector may exceed the layer thickness and determine the spectrum shape. At oblique incidence and backscattering the projected path in the layer is much wider than the minimum depth resolution of the detector.

5.6 GLANCING INCIDENCE RUTHERFORD BACKSCATTERING

A comparison of the two geometries, normal incidence and glancing incidence, is shown in Fig. 5.9 (Williams and Grant, 1974). For a thin implant layer we are limited by the detector resolution to some 200 Å. By entering and leaving at glancing incidence we are using the backscattering analogue of taking a taper section, lengthening the path in the thin layer and greatly improving the resolution. A limitation of any low angle of incidence method is surface finish. The problem has been discussed by Schmid and Ryssel (1974) and Williams (1975). Another limitation is the onset of multiple scattering. Rutherford backscattering usually relies on a single large angle scattering event which means that the depth probed is limited to about a "mean free path" of the projectile in the materials so that as we increase the obliqueness of incidence and the resolution we decrease

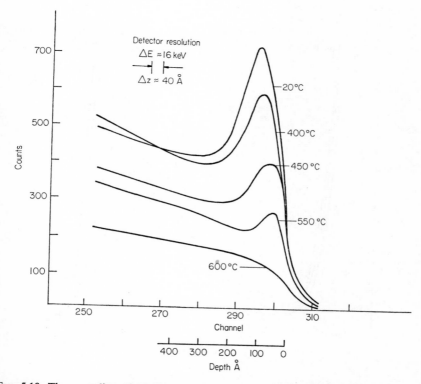

FIG. 5.10. The annealing of implanted 20 keV Dy^+ implanted into Ni at 20°C. The as implanted peak is 110 Å below the surface and moves towards the surface as the annealing temperature is increased, showing that the damage anneals from the undamaged region outwards.

the depth that we can probe. Again this is not a serious limitation for many systems as high resolution is most necessary for thin near surface layers.

A useful example of the application of this high resolution technique to annealing has been made by Stephens *et al.* (1974) for isochronal annealing of 20 keV Dy$^+$ implanted at 20°C into Ni. The system was chosen because of the interest in rare earth ferromagnetic alloys. Figure 5.10 shows the annealing results using 2 MeV He$^+$ glancing incidence channelling spectra $\phi = 80°$ and $\langle 110 \rangle$ which gives a depth resolution in this system of 40 Å, using a detector with $\Delta E = 16$ keV. The unannealed damage peaks at 110 Å

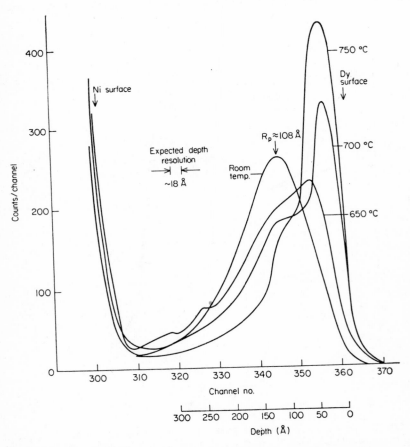

FIG. 5.11. Random spectra at 84° to the surface normal for Dy implanted into single crystal Ni, showing migration of Dy towards the surface at over 600°C.

below the surface with a tail extending to 270 Å. The curves clearly show that the damage peak moves towards the surface during annealing, that is, the crystal is reordering itself from the undamaged region. This is not unexpected and has been observed for example in the annealing of the irradiation induced amorphous layer on silicon (Mayer et al., 1968). The damage, as measured by the area under the disorder peak, anneals out in the range 300–600°C, with little change occurring below 300°C, and little change in the Dy implant profile. Above 600°C the Dy starts to move, diffusing out and accumulating near the Ni surface. Figure 5.11 shows Dy spectra taken in a random direction, at $\phi = 84°$ which gives a depth resolution of 10 Å. The change from a near Gaussian profile at room temperature to a skew profile strongly peaked near the surface after annealing is apparent.

Another example of the glancing incidence technique has been to study the change in implant profile with dose (Williams and Grant, 1974). The depth profiles of Pb, implanted at room temperature in Si at 20 keV for various doses from 5×10^{14} to 5×10^{16} cm^{-2} are shown in Fig. 5.12. At the lower doses the profile is nearly Gaussian, but is starting to flatten out at 5×10^{15} cm^{-2} with thereafter an accumulation of Pb near the Si surface. The spectrum shape is probably due to the formation of small precipitates of lead in regions where the local solubility limit has been exceeded. We thus have two groups of backscattered particles, He reflected from small areas of Pb, and He reflected from the usual kind of Pb implant distribution dispersed in the Si lattice, leading to the double humped spectrum in Fig. 5.12. In the calculation of a depth scale the extra stopping power of the Pb must be considered and if some of it is dispersed as small particles, their size and distribution will influence the depth scale.

Excellent depth resolution has been previously obtained by the use of magnetic or electrostatic analysers, for example a resolution of several angstrom units for 80 keV protons on gold and nickel films by van Wijngaarden et al. (1971). Surface barrier detectors respond to both charged and neutral backscattered particles and with a multichannel analyser count every particle. Electrostatic or magnetic analysers count only a particular energy window at a time and as they only respond to ions they require calibration and frequently one has more particles reflected as neutrals than as ions. Special cooled solid state detectors with an energy resolution of 2–3 kV have also been used by van der Weg et al. (1973), to obtain higher depth resolution without recourse to glancing incidence. These and similar methods are discussed in a number of recent review articles, for example Buck and Poate (1974).

It is clear that such high resolution methods will be brought to bear on the problems of diffusion, oxidation, corrosion and stoichiometric variations that occur over small distances near layer interfaces.

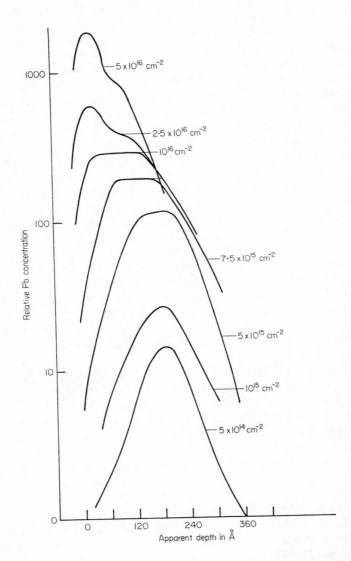

FIG. 5.12. The range distribution of 20 keV Pb$^+$ implanted at room temperature into Si as a function of dose, from 5×10^{14} to 5×10^{16} cm^{-2}. Above a dose of 5×10^{15} the original Gaussian profile flattens and then develops a double peak structure, believed to be associated with precipitation of lead near the surface.

5.7 INTERSTITIAL DIFFUSION

When implants are made in a channelling direction the channelled component, suffering fewer large angle collisions, will penetrate to a greater depth. However, a small proportion of the ions travel even further (Domeij et al., 1964). This supertail has been shown by Davies and Jespersgård (1966) to be due to rapid interstitial diffusion. They implanted ^{42}K at 40 keV along the $\langle 100 \rangle$ direction into tungsten at 30 K. The results are shown in the integral penetration curves of Fig. 5.13. If the specimen was warmed to room temperature the supertail appeared. If, however a further heavy bombardment with Ne was given after the K implant and before raising the temperature no supertail appeared.

The results indicate interstitial diffusion, where the channelled implant comes gently to rest in an interstitial position without creating much damage at the end of its range. The atom is then free to diffuse considerable distances before it is trapped by a defect. The bombardment after implantation and before warming creates defects near the end of the range of the channelled implants into which they fall without diffusing to depths of several microns

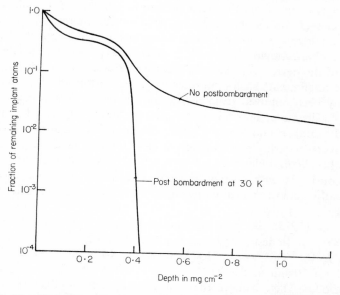

FIG. 5.13. The effect of trapping centres produced by postbombardment on the interstitial diffusion of 40 keV implanted ^{42}K into W along the $\langle 100 \rangle$ direction at 30 K. The upper curve shows a supertail in the absence of the postbombardment. The lower curve shows the effect of a postbombardment of 3×10^{15} Ne cm^{-2} before warming the crystal. Both curves are for the crystal warmed to room temperature.

as occurs in the unpostbombarded case. If the supertail were due to a channelling effect and not a subsequent diffusion, there would be little difference in the profiles in the two cases. The activation energy for diffusion deduced from these measurements was 0·5–0·8 eV compared with 4–5 eV for normal bulk diffusion, lending further support to the interstitial diffusion mechanism.

A number of other systems in which interstitial diffusion is believed to operate are given by Mayer *et al.* (1970). In general the mechanism will only be important, (McCaldin, 1965), for small damage at the end of the implant atom's track and this in general means $Z_1 < Z_2$; for implant atoms which do not favour interchange with the host lattice atoms or for low activation energies for interstitial diffusion.

The ion distributions have been calculated by Sparks (1969), using multi-stream steady state diffusion models which give reasonable agreement with the experimental results.

5.8 ION IMPLANTATION GETTERING OF DAMAGE

Gettering in vacuum tubes means evaporating an active film on the glass walls after the tube is evacuated and sealed. The film reacts with any residual gas which may evolve from the electrodes during the life of the tube and is the forerunner of the modern titanium sublimation pumps. In solid state devices an analogous situation occurs where heavy metal impurities in the diode space charge region can have a deleterious effect on the electrical behaviour of the device, for example excessive reverse leakage current. It is not feasible to eliminate such impurities from the starting material and they have usually been removed from the active region by treating the surface with phosphosilicate or borosilicate glass layers and giving a high temperature anneal (Lambert and Reese (1968)). Mechanical damage has also been found to assist gettering and Nakamura *et al.* (1968) have suggested that defects in the silicon under the glass play an important part in this process, which is consistent with current models of diffusion in damaged solids. More recently quantitative attempts have been made to characterise the diffusion processes involved.

Hsieh *et al.* (1973), using silicon photodiode array camera targets, each with 800,000 p^+n diodes, found that bombarding with 10^{16} ions per cm^{-2} P at 50 keV, 10^{15} cm^{-2} As at 150 keV or 3×10^{15} cm^{-2} Ar at 50 keV, followed by a 30 min anneal at 900°C in dry N_2 with 5% O_2, eliminated defective diodes. They deduced that ion irradiation is an effective gettering "dark field" treatment and that it is the irradiation damage which is important and not the ion species used. Buck *et al.* (1973) have applied Rutherford backscattering to the problem and found that Cu and Ni getter rapidly and Fe, Co and Au move slowly in ion damaged Si. Their experiment consisted of evaporating a thin metal film on one side of a 125 μm Si wafer, the other side of which had been irradiated with 10^{16} ions per cm^{-2} of 100 keV Si^+. After 900°C anneals for 30 min Rutherford backscattering at 180° was used to measure the diffusion profiles of damaged and undamaged areas. All of the metals studied diffuse interstitially in Si but at thermal equilibrium Cu and Ni (the fast diffusers) are mainly interstitial and the other three mainly substitutional, which suggests that gettering is limited by the diffusivity and interstitial concentration of the impurity metal.

There are other more complex effects such as greatly enhanced diffusions in heavily damaged Si of both Cu and O_2, when both are present, which greatly exceeds the diffusion of either Cu or O_2 alone (Poate and Seidel (1973)).

The results to date should serve as a warning that in experiments where ion bombardment cleaning and heat treatments are used, the distribution of near surface atoms may be effected.

5.9 THE DIFFUSION OF HELIUM IN COPPER

The motion of gases in metals has been of interest particularly for reactor materials where considerable numbers of energetic noble gas atoms are amongst the fission products. This work is discussed briefly in Section 10.16 and the implications for the structural integrity of the fuel cans etc. are considered. We will mention here the experimentally more difficult case of helium and confine ourselves to some recent backscattering measurements of its diffusion in copper. The cross section for Rutherford backscattering, varying as it does as the square of the atomic number of the target atom, gives a helium yield 210 times less than the host copper atoms in which the helium

TABLE 5.1

Projected range R_p and range straggling ΔR_p for He implanted into Cu films

Implant		R_p (Å)		ΔR_p (Å)	
E (KeV)	Fluence ($\times 10^{17} cm^{-2}$)	Measured	Calculated	Measured	Calculated
54·0	3	1670	1700	590	720
104·5	1	2960	3050	660	960
157·6	1	4340	4280	620	1110

is implanted. For a thick Cu target the result will be a He peak which is much more difficult to detect than the O_2 peak shown in Fig. 5.7. The solution is to remove the Cu spectrum by using a thin Cu film. The result is a spectrum rather like Fig. 5.2 but without the Be edge, and with a He peak in the mass 4 position. Blewer (1973) has used this method to attain a detection sensitivity better than one atomic % for He implanted into polycrystalline Cu at 50, 100 and 150 keV. The peaks of the helium distributions were located to $\pm$ 100 Å using 2·5 MeV protons backscattered at 164° and a 10 keV resolution detector, a resolution which could be improved by backscattering at grazing incidence. Blewer's results for projected range, R_p, and range straggling, ΔR_p, are compared with the calculated values (Brice, 1971) in Table 5.1.

The ranges agree well with the calculated values, within 3%, but the straggling is considerably smaller than predicted, 44% less for the highest energy implants. The peaks were symmetric Gaussian distributions and showed no evidence of channelling or interstitial diffusion.

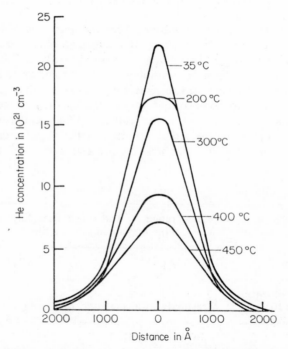

FIG. 5.14. Change in concentration of He implanted into polycrystalline copper foil under 35 min isochronal anneals at temperatures up to 450°C. At 200°C enhanced diffusion in the most damaged region is apparent. At higher temperatures the helium is lost symmetrically to both sides of the peak.

The measured profile of the He after 35 minute isochronal anneals is shown in Fig. 5.14 (Blewer, 1974). A flattening of the profile at 200°C is probably due to enhanced diffusion caused by damage at the end of the range of the implanted atoms. At higher temperatures the profile flattens out and shows a loss of He from the foil. Scanning electron microscope studies of the foils show that bubbles form and rupture as the temperature is increased, the rate determining step being failure of the metal containing the bubble as the internal pressure increases with temperature. Mass spectrometric measurements showed that the He disappearing from the profiles of Fig. 5.14 was evolved.

These observations, and similar ones, indicate that the motion of gases through and out of solids is a much more complex process than is indicated by assigning a diffusion coefficient and an activation energy alone (e.g. Kornelsen, 1972).

REFERENCES

Amsel, G. (1973). ENS Report No. 10.
Blewer, R. S. (1973). *Appl. Phys. Lett.* **23**, 593.
Blewer, R. S. (1974). *In* "Applications of Ion Beams to Metals" (Eds S. T. Picraux, E. P. Eer Nisse and F. L. Vook), Plenum Press, pp. 557.
Boltaks, B. I. (1963). "Diffusion in Semiconductors", Academic Press, New York.
Bøgh, E. (1968). *Can. J. Phys.* **46**, 653.
Borders, J. A. (1973). *Thin Solid Films*, **19**, 359.
Bourgoin, J. and Corbett, J. W. (1972). *Phys. Lett.* **38A**, 135.
Brice, D. K. (1971). *Rad. Eff.* **11**, 227.
Buck, T. M. and Poate, J. M. (1974). *J. Vac. Sci.* **11**, 289.
Buck, T. M., Poate, J. M., Pickar, K. A. and Hsieh, C. M. (1973). *Surface Sci.* **35**, 362.
Cookson, J. A., Ferguson, A. T. G. and Pilling, F. D. (1972). *J. Radioanalytical Chem.* **12**, 39.
Davies, J. A. and Jespersgård, P. (1966). *Can. J. Phys.* **44**, 1631.
Dearnaley, G. and Hartley, N. E. W. (1974). *Phys. Lett.* **46A**, 345.
DeBonte, W. J., Poate, J. M., Melliar-Smith, C. M. and Levesque, R. A., 1973. "Proc. Albuquerque Meeting on Interaction of Ion Beams With Metals".
Domeij, B., Brown, F., Davies, J. A., Piercy, G. R. and Kornelsen, E. V. (1964). *Phys. Rev. Letters*, **12**, 363.
Eer Nisse, E. P. (1971). *Appl. Phys. Lett.* **18**, 581.
Eer Nisse, E. P. (1972). *J. Appl. Phys.* **43**, 1330.
Eer Nisse, E. P. (1974). *In* "Ion Implantation in Semiconductors and Other Materials" (Ed. B. L. Crowder), Plenum, New York, pp. 531.
English, A. T. and Turner, P. A. (1972). *J. Electronic Materials*, **1**, 1.
Gertsniken, S. D. and Dekhtyar, I. Ya (1960). "Solid State Diffusion in Metals and Alloys", USAEC translation 6313.
Hartley, N. E. W. (1975). *J. Vac. Sci. Technol.* **12**, 485.
Hsieh, C. M., Mathews, J. R., Seidel, H. D., Pikar, K. A. and Drum, C. M. (1973). *Appl. Phys. Lett.* **22**, 238.

Kornelsen, E. V. (1972). *Rad. Eff.* **13,** 227.
Lambert, J. L. and Reese, M. (1968). *Solid State Electronics,* **11,** 1055.
LeClaire, A. D. (1949). *Prog. in Metal Physics,* **1,** 306.
McCaldin, J. O. (1965). *Prog. Solid State Chem.* **2,** 9.
Mayer, J. W., Eriksson, L. and Davies, J. A. (1970). "Ion Implantation in Semiconductors", Academic Press.
Mayer, J. W., Eriksson, L., Picraux, S. T. and Davies, J. A. (1968). *Can. J. Phys.* **46,** 663.
Morgan, D. V. (1974). *J. Phys.* D, **7,** 653.
Myers, S. M. (1974). *J. Appl. Phys.* **45,** 4320.
Myers, S. M., Picraux, S. T. and Prevender, T. S. (1974). *Phys. Rev.* **B9,** 3953.
Nakamura, M., Kato, T. and Oi, N. (1968). *Japan J. Appl. Phys.* **7,** 512.
Poate, J. M. and Seidel, T. E. (1973). *In* "Ion Implantation in Semiconductors and Other Materials" (Ed. B. L. Crowder) p. 317, Plenum Press, New York.
Primak, W. (1969). *Surface Sci.* **16,** 398.
Schmid, K. and Ryssel, H. (1974). *Nucl. Inst. and Meth.* **119,** 287.
Sippel, R. F. (1959). *Phys. Rev.* **115,** 1441.
Sparks, M. (1969). *Phys. Rev.* **184,** 416.
Stephens, G. A., Robinson, E. and Williams, J. S. (1974). Conf. on "Ion Implantation in Semiconductors and Other Materials", Osaka.
van Bueren, H. G. (1961). "Imperfections in Crystals", North-Holland, Amsterdam.
van der Weg, W. F., Kool, W. H., Roosendahl, H. E. and Saris, F. W. (1973). *Rad. Eff.* **17,** 245.
Van Wijngaarden, A., Miremadi, B. and Baylis, W. E. (1971). *Can. J. Phys.* **49,** 2440.
Williams, J. S. (1974). *Rad. Effects,* **22,** 211.
Williams, J. S. (1975). To be published.
Williams, J. S. and Grant, W. A. (1974). Conf. on "Ion Implantation in Semiconductors and Other Materials", Osaka.
Williams, J. S. and Kelly, J. C. (1975). To be published.

CHAPTER 6

Sputtering

6.1 INTRODUCTION

During ion bombardment of a target energy can be deposited near the surface at such a high rate that atoms are ejected from the solid. The measured sputtering yield is normally in the range 1 to 50 target atoms removed per incident ion so that for typical implantation doses the surface is moved by some tens of monolayers. However, if we consciously alter the energy and mass of the bombarding ion so that the energy is deposited within a 100 Å of the surface then the sputtering ceases to be an annoying adjunct of ion implantation and instead becomes an extremely powerful tool for machining, contouring or polishing the surface.

The inherent advantages of an ion beam for machining are (i) the beam definition which can form features less than 5000 Å wide and in depth may control to 50 Å. (ii) The shaping "tool" is an ion beam so it does not deform or wear during the machining and there is a complete absence of lubricating or coolant fluids. (iii) The sputtering proceeds on an atomic scale so it is equally applicable to metals, semiconductors or insulators and may be performed at any stage in the preparation of the material.

The technological opportunities for such a machining process are enormous with obvious examples in the machining of glass, aspherics, contoured surfaces, composite or prefabricated systems. In conjunction with the more rapid conventional techniques the ion beam machining may provide final tolerance control, dimensional control after case hardening, or remove the strain region on the surface of a solid which has been polished by normal means. Numerous applications of sputtering will be mentioned in both this chapter and Chapter 9. Besides offering new dimensions to machining techniques one should also consider if ion beam machining is economically viable. Fortunately it appears that the costs in time and money are similar to conventional processes in the cases where comparisons have been made.

Sputtering is a function of many variables including the masses of the ion and the target atom, the ion energy, direction of incidence to the face of the target, the target temperature and the ion flux (current density). The measured sputtering yield will also depend on the background gas pressure, the previous history of the solid, including the concentration of implanted ions, and electrical fields existing at the surface of the target. Not all measurements of the sputtering yield have been made under ideal conditions so numbers quoted in the literature for the same ion-target combination may disagree. In general, recent high vacuum results may be reproducible, but these yields can only offer a guide to the actual sputtering yields obtained in a particular set of apparatus.

The changes in sputtering yield with (M_1, M_2, E, θ, T) are in accord with theoretical predictions for both amorphous and crystalline solids. It may be convenient to note that for technological applications one may ignore the theoretical description of sputtering and just use the empirical values of the sputtering yields. However, it is essential to realise that practical materials are rarely featureless, nor are they completely isotropic. Ion beam sputtering will reveal both types of deficiency and will generate new surface features. We will therefore discuss topographical effects in some detail.

Additional books or review articles which discuss sputtering, surface topography and applications include: Wehner (1957), Kaminsky (1965), Carter and Colligon (1968), MacDonald (1970), Navinsek (1972), Sigmund (1972, a, b), Behrisch et al. (1973), Norgate and Hammond (1974), Wilson (1974), McCracken (1975).

6.2 SPUTTERING THEORY

In order to explain why many target atoms are sputtered per incident ion we require a sputtering theory which couples in the energy deposited over a small volume of the solid and transposes it to the surface layer. The alternative sources of sputtering are shown in Fig. 6.1. The larger volume of Fig. 6.1(b) was thought to result from either a thermal spike or a collision cascade. The predictions of a thermal spike, with a high local "temperature" were unsatisfactory and the collision cascade theory of sputtering adequately predicts the sputtering rate and the energy distribution of the ejected material. If the solid is crystalline then some atoms will be ejected in crystal directions as a result of replacement collision sequences and focussing effects of the lattice. Whilst they alter the angular pattern of the ejected material such sequences do not replace the collision cascade model. Indeed the range of such sequences is still in dispute.

A typical cascade has a radius of 100Å and energy is dissipated both by elastic displacements and electronic loss. The total number of displacements will be a function of the state of excitation of the region so if, for example, the atoms have already been excited by one cascade then a second cascade will produce more displacements and sputtering. At normal current densities

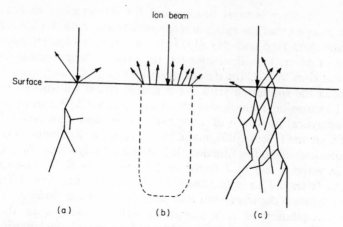

FIG. 6.1. (a) Sputtering of surface atoms resulting from an interaction between the incident ion and the surface layer. (b) Evaporation from a thermal spike. (c) Sputtering from energy directed toward the surface as a result of a collision cascade.

the residual damage regions of cascades will overlap but there is only a small probability that the regions are simultaneously excited by both. If each collision event requires 10^{-14}s then energy will be transferred across the 100 Å radius of a cascade in $\sim 10^{-11}$ seconds.

However, we can artificially produce a correlation between cascades by bombardment with molecules. These will dissociate at the surface and the constituent ions will form overlapping cascades. The sputtering yield is then

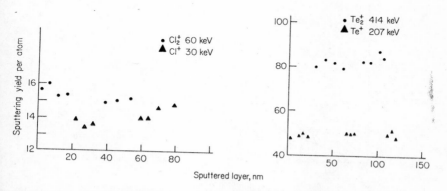

FIG. 6.2. Sputtering yields from silver as a function of alternate bombardments with ions and molecular ions. The overlap of collision cascades changes the sputtering yield and is more effective for the tellurium ions than the chlorine ions ($M_{Te} = 128$, $M_{Cl} = 35$ a.m.u.).

enhanced. The effect is most pronounced for intermediate to heavy projectiles at energies near the maximum for nuclear stopping. Figure 6.2 shows results from Andersen and Bay (1973) for alternate molecular and atomic sputtering of silver. For the lighter ion the average enhancement in rate is only 9% but rises by 67% for the tellurium ions.

To predict the sputtering yield is not simple and if we wish to derive an analytical expression many assumptions are necessary. We will commence with an analytical approach of Thompson (1968) and then proceed to the theory of Sigmund (1969a,b). Sigmund's theory requires a computer evaluation of the damage distribution function. It is quantitatively successful for $M_2 \leqslant M_1$ but in error by up to a factor of 2 when $M_2 \sim 10 M_1$. This is quite acceptable. It should be emphasised that the "real" situation differs from the theory because the implanted ions alter the substrate material. This can produce precipitation of new phases as well as changes in the yield. Measured sputtering yields are therefore a function of implantation dose. In hindsight the early measurements of Almén and Bruce (1961), which show a Z_1 dependence, should have been termed the sputtering yield of the "alloy" $(Z_1 + Z_2)$. Their results together with an example of Sigmund's prediction are shown in Fig. 6.3; (Andersen and Bay 1972a, b).

Examples of the change in sputtering yield with ion dose are shown in Fig. 6.4 for vanadium and bismuth ions of 45 keV energy incident of copper. At low implant doses ($< 10^{16}$ ions cm^{-2}) the yield is accurately described

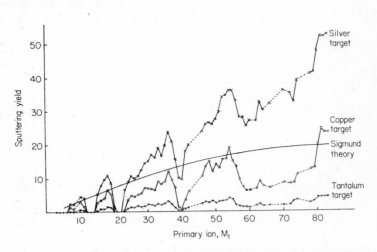

FIG. 6.3. The Z_1 dependence of saturation sputtering yields (i.e. target is an alloy of $Z_1 Z_2$) for 45 keV ions. Sigmund's calculated yield for copper is also shown.

by Sigmund's theory but at higher doses the yields diverge from the predicted values. The data of Almén and Bruce (1961) is in good agreement with these experiments of Andersen (1973).

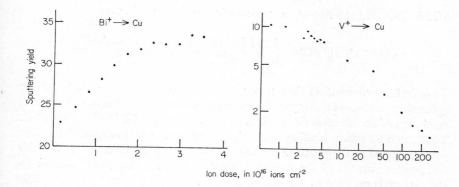

FIG. 6.4. The dose dependence of the sputtering yield of copper for bombardment with 45 keV ions. Note the difference in scales for the Bi$^+$ and V$^+$ data.

6.3 THE THOMPSON MODEL OF SPUTTERING

Thompson (1968) considered two body collisions in an amorphous solid in which the number of primary ions formed at energy E_2 was $q(E_2)\,dE_2$ per unit volume per second. They transfer energy E' to secondary ions with an efficiency $v(E_2, E')$ so the total number of ions slowing down by E' is

$$\int_{E'}^{\infty} q(E_2)\,v(E_2, E')\,dE_2.$$

The function $v(E_2, E')$ is of course the radiation damage function (see Section 4.5) and is approximately $\eta E_2/E_1$, where $\eta \sim 1$ or 2.

If the secondary atom has a velocity vector $\mathbf{V}'$ then the energy loss is

$$\frac{dE'}{dt} = \mathbf{V}' \frac{dE'}{dx} \quad \text{or} \quad dt = \frac{dE'}{V'(dE'/dx)}.$$

The density of atoms $\rho\,(E', \mathbf{r})\,dE'\,d\Omega'$ in the energy range E' to $E' + dE'$ in a solid angle $d\Omega'$ about the vector direction $\mathbf{r}$ is then

$$\rho(E'\,\mathbf{r})\,dE'\,d\Omega' = \int_{E'}^{\infty} q(E_2)\,v(E_2, E')\,dE_2 \frac{d\Omega'}{4\pi}\,dt.$$

The flux of ions crossing a surface whose normal is at an angle θ to $\mathbf{r}'$ is

$$\Phi'(E', \mathbf{r}')\,dE'\,d\Omega' = \mathbf{V}'\,\rho(E', \mathbf{r}')\cos\theta\,d\Omega'\,dE'.$$

Within the solid the flux of atoms crossing an internal surface is

$$\Phi'(E', \mathbf{r}')\,dE'\,d\Omega' = \int_{E'}^{\infty} \frac{q(E_2)\,v(E_2, E')}{dE'/dx}\,dE_2\,\frac{\cos\theta\,d\Omega'\,dE'}{4\pi}.$$

To a first approximation the surface is like a plane within the solid and the sputtering yield can be computed by integrating over a hemisphere. Assuming rigid ions of an interatomic spacing, D, the energy loss dE'/dx is about E'/D. To include a surface binding term, E_B, Thompson (1968) suggested the ions were refracted as they left the surface. Refraction occurs because the velocity component parallel to the surface is unaltered whilst the energy normal to the surface is reduced by the binding energy. Choosing θ_1 and ϕ as the angle between the trajectories and the normal inside and outside the surfaces gives

$$V'\sin\theta_1 = V\sin\phi$$

$$\tfrac{1}{2}M_2 V'^2 \cos^2\theta_1 = \tfrac{1}{2}M_2 V^2 \cos^2\phi + E_B$$

and
$$E' = E + E_B.$$

We have made the assumption that E_B is not directional. The internal solid angle, $d\Omega' = 2\pi\sin\theta_1\,d\theta_1$, so as we progress across the surface the flux is constant if the internal flux (prime quantities) and ejected flux are related by $\Phi'(E', \theta)\,dE'\sin\theta_1\,d\theta_1 = \Phi(E, \phi)\,dE\sin\phi\,d\phi$. We also note

$$\frac{\sin\theta_1}{\sin\phi} = \frac{1}{\sqrt{(1 + E_B/E)}}.$$

This leads to an ejected flux

$$\Phi(E, \phi)\,d\Omega\,dE = \frac{\eta D\cos\phi}{4\pi(1 + E_B/E)^3}\,\frac{1}{E^2}\int_{E+E_B}^{\infty} E_2\,q(E_2)\,dE_2\,d\Omega\,dE.$$

At any particular angle $\Phi(E) \propto 1/E^2$ if $E_B \leqslant E \leqslant E_2$. Which is close to the observed energy dependence. To evaluate the sputtering yield Thompson (1968) specified $q(E_2)$ via an inverse square potential

$$V(r) = \frac{2E_R}{e}(Z_1 Z_2)^{5/6}\left(\frac{a_0}{r}\right)^2$$

(where E_R = the Rydberg energy, $a_0 = 0.53$ Å) this gives

$$q(E_2) = \begin{cases} 0 & \text{if } E_2 > \dfrac{4 M_1 M_2 E_1}{(M_1 + M_2)^2} \\ \dfrac{\pi^2 a_0^2 N E_R (Z_1 Z_2)^{5/6}}{2 e E_1^{1/2} E_2^{3/2}} \left(\dfrac{M_1}{M_2}\right)^{1/2} \Phi_1 & \text{if } E_2 < \dfrac{4 M_1 M_2 E_1}{(M_1 + M_2)^2} \end{cases}$$

Φ_1 is the flux of ions crossing unit area normal to the path and N is the density of atoms.

If the ion beam is incident at an angle θ then the sputtering yield per incident ion will be

$$S = \frac{\Phi(E, \phi) \, d\Omega \, dE}{\Phi_1 \cos \theta}.$$

For the potential and approximation chosen this predicts

$$S = \frac{\pi^2 a_0^2 N^{2/3}}{8e} \frac{E_R}{E_B} \frac{M_1 (Z_1 Z_2)^{5/6}}{(M_1 + M_2)} \sec \theta.$$

The sec θ dependence is observed for $\theta \lesssim 60°$ for $M_2 \gg M_1$ which, of course, is the case where the inverse square potential is reasonably accurate. Thompson's (1968) predicted sputtering yields were compared with the data of Almén and Bruce (1961), which in hindsight (Section 6.2) was inappropriate for most of the targets quoted. He chose E_B values from Honig (1962) and calculated the sputtering yield as 15, for copper during bombardment with 45 keV ions, a more recent measurement agrees with this (Andersen and Bay, 1972a, b).

6.4 THE SIGMUND MODEL OF SPUTTERING

In the Sigmund (1969) treatment of a collision cascade he considers the energy transfer from a primary ion to the atoms of an isotropic solid. The "surface" of the solid is defined by a plane in this continuous medium at which the primary ion is injected. We saw in Section 2.13 that the damage distribution function which resulted from this (Figs 2.21, 2.22, 6.5) includes a region outside of the surface, that is, the shaded region of Fig. 6.5. For energy to have reached this region we must assume that either, the primary ion was reflected, or target atoms were "sputtered". Clearly the isotropic solid model is not perfect because we have not yet included a surface binding energy, nor rejected atoms which re-enter the solid. In essence we shall be looking for an expression for the sputtering yield which depends on the

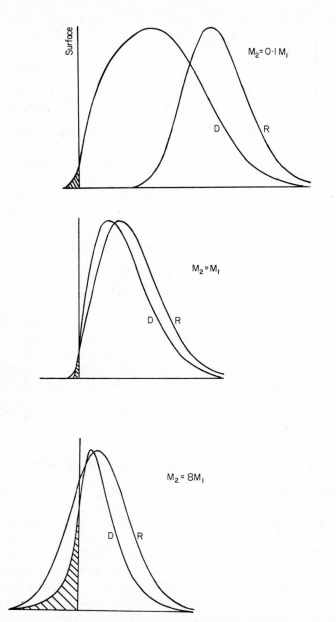

FIG. 6.5. A comparison of the calculated damage and implant profiles for (a) $M_2 = 0·1\,M_1$; (b) $M_2 = M_1$; (c) $M_2 = 8\,M_1$. The regions to the left of the "surface" represent sputtering for the damage calculation and reflection for the ion range estimate.

6.4 THE SIGMUND MODEL OF SPUTTERING

stopping cross section for displacement, $S_n(E)$, the efficiency of energy transfer, $\alpha(M_2/M_1)$, and a surface binding, U_0 (in practice we use the sublimation energy).

Hence the sputtering yield $S = (\lambda S_n(E)\alpha(M_2/M_1))/U_0$ where λ should be a calculable constant.

As a first step to estimate the sputtering we can define a sputtering efficiency, γ, as

$$\gamma = \frac{1}{E}\int_{-\infty}^{0} F(x)\,dx$$

where $F(x)$ is the distribution function of damage with depth below the surface. The second step of actually deriving the distribution function is much more difficult. Sigmund (1969, 1972a, b) proceeds by writing a Boltzmann transport equation for the atoms, commencing with a primary energy E. If we label the atom velocity by V, energy E_0 and transferred energy T at some time t then the equation is

$$-\frac{1}{V}\frac{\partial}{\partial t}F(E, E_0, t) = N\int d\sigma\,\{F(E, E_0, t) - F(E - T, E_0, t) - F(T, E_0, t)\}$$

$$+ N S_e(E)\frac{\partial}{\partial E}F(E, E_0, t)$$

where N is the atomic density and the electronic and nuclear stopping cross sections are related to the stopping power by

$$\left.\frac{dE}{dx}\right)_n = -N S_n(E) \quad \text{and} \quad S_n(E) = \int_0^{T\,\text{max}} T\,d\sigma.$$

To remove the time dependence of this function, F, we define a second function G which is the time averaged distribution function of the energy of the moving atoms and

$$G(E, E_0)\,dE_0 = \psi\,dE_0 \int_0^{\infty} dt\,F(E, E_0, t).$$

Here ψ is the flux of incident ions. To evaluate G we can write

$$\frac{\psi}{V_0}\delta(E - E_0) = N\int d\sigma\,\{G(E, E_0) - G(E - T, E_0) - G(T, E_0)\}$$

$$+ N S_e\frac{\partial}{\partial E}G(E, E_0).$$

Sigmund (1969) rearranges this expression in terms of Legendre polynomials and uses the method of moments to obtain a solution. To include the energy dependence of the nuclear and electronic scattering cross sections he uses the power potential approximations of the Thomas–Fermi atomic model. This gives a solution for the differential cross section for energy

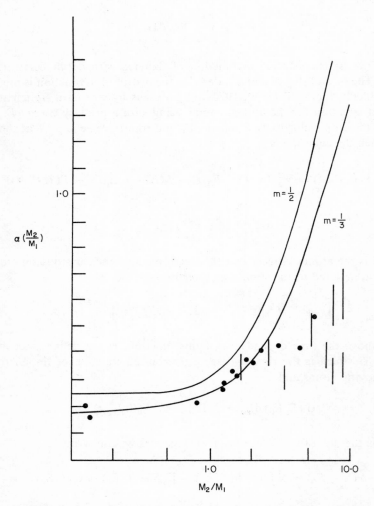

FIG. 6.6. Two calculations of the factor $\alpha(M_2/M_1)$ for power law expressions with $m = \frac{1}{2}$ and $m = \frac{1}{3}$ compared with experimental data for ions incident on silicon, copper and silver.

6.4 THE SIGMUND MODEL OF SPUTTERING

transfer T as

$$d\sigma(T) = C E^{-m} T^{1-m} dT$$

where

$$C = \frac{\pi \lambda_m a^2}{2} \left(\frac{M_1}{M_2}\right)^m \left(\frac{2 Z_1 Z_2 e^2}{a}\right)^{2m}.$$

The exponent $(1/m)$ in the power potential is chosen as $m = 1$ at high energies (Rutherford scattering), $m = \frac{1}{2}$ for medium energies ($\sim$10 to 100 keV), $m = \frac{1}{3}$ ($\sim$ 10 to 1000 eV) and $m = 0$ at very low energies. Appropriate values for λ_m are $\lambda_1 = 0.5$; $\lambda_{1/3} = 1.309$; $\lambda_{1/2} = 0.327$; $\lambda_0 = 24$.

Surface binding is introduced by means of a spherical barrier so that the potential is equally effective for ions leaving the surface in any direction. The most important consequence of the surface binding energy is that it influences the energy spectrum of the ejected atoms. Because ejected atoms are of low energy Sigmund (1969) used $m = 0$, and calculated an energy distribution proportional to E_0^{-2} (for other values of m it becomes $E_0^{(2m-2)}$). His prediction of the sputtering yield, S, (average number of sputtered atoms per incident ion) followed as

$$S = \frac{3}{4\pi^2} \frac{F(E)}{N C_0 U_0}.$$

The target material determines N, C_0 and U_0 and the incident ion enters the expression via $F(E)$ because the damage distribution function is controlled

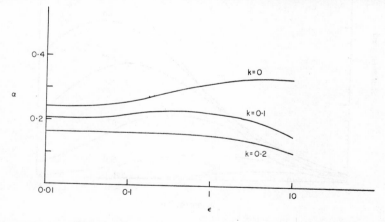

FIG. 6.7. The variation of the factor α for $M_1 = M_2$ with energy, in reduced energy units, including electronic stopping, calculated for $m = \frac{1}{3}$.

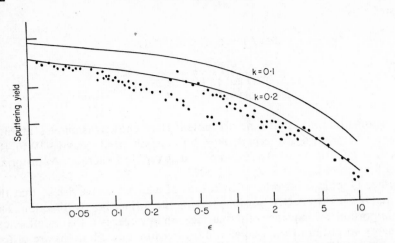

FIG. 6.8. A comparison of experimental and calculated sputtering efficiences for a range of energies and ion-target combinations including Kr–Ag, Xe–Pb, Ar–Cu and Ne–Si data. The experimental mass ratio is $M_2/M_1 = 1.6$. The theoretical curves show the effect of the effect of the electronic energy loss for equal masses.

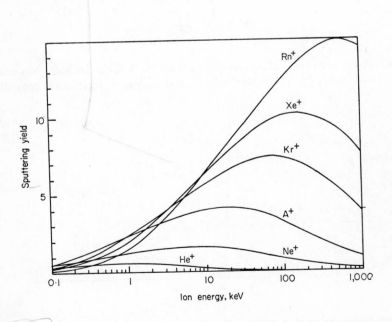

FIG. 6.9. Calculated sputtering yields for aluminium as a function of ion energy and ion mass.

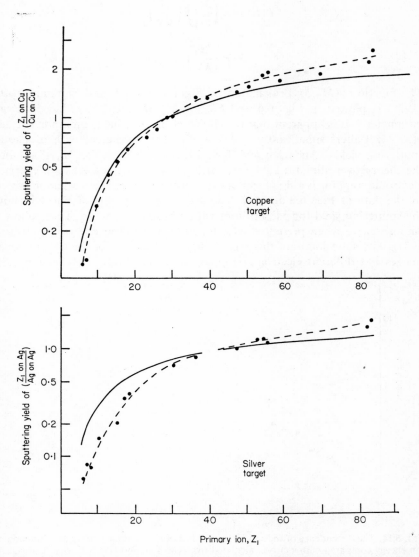

FIG. 6.10. The relative sputtering yields of polycrystalline copper and silver targets obtained by bombardment with 45 keV ions. The data is normalised at the self sputtering yield and is compared with the theoretical prediction of Sigmund.

by the energy dependence of the stopping power and the relative masses of the target and incident ions. If we separate these contributions

$$F(E) = \alpha \left(\frac{M_2}{M_1}\right) \left(\frac{dE}{dx}\right)_n$$

$$= \alpha \left(\frac{M_2}{M_1}\right) N S_n(E).$$

The function $\alpha(M_2/M_1)$ calculated by Sigmund (1969) and Sigmund *et al.* (1971) is plotted in Figs 6.6 and 6.7 for M_2/M_1 and the reduced energy parameter ε. It is apparent that the choice of the m value is unimportant but $\alpha(M_2/M_1)$ alters significantly if $M_2 > M_1$. The agreement with measured sputtering yields (Andersen and Bay, 1973) is excellent for $M_2 \leqslant M_1$ but the theory overestimates $\alpha(M_2/M_1)$ at high target masses. At high energies electronic stopping is a significant form of energy loss and this is demonstrated by the data of Figs 6.8 and 6.9. The experimental points of Fig. 6.8 show the sputtering yield for pairs of ions with a mass ratio of 1·6. They follow a smooth curve when presented as a function of reduced energy (Andersen, 1971). The solid lines are theoretical curves calculated for the case of equal masses but different electronic stopping (i.e. the k values of Section 2.5).

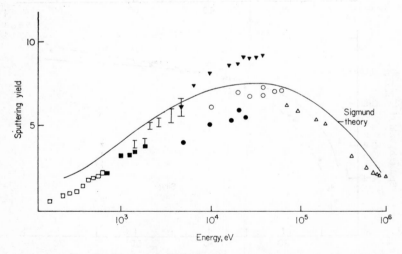

FIG. 6.11. The dependence of the sputtering yield of polycrystalline copper as a function of ion energy compared with a curve calculated from the Sigmund (1969) theory. The experimental data is for argon bombardment and taken from the work of: Almén *et al.* (1961), (○○○); Dupp *et al.* (1966), (△△△); Guseva (1960), (●●●); Southern *et al.* (1963), (III); Wehner *et al.* (1961), (□□□); Weijsenfeld (1967), (■■■); and Yonts *et al.* (1960), (▼▼▼).

6.4 THE SIGMUND MODEL OF SPUTTERING

An alternative presentation of this type of data is given by Fig. 6.9 which gives the calculated sputtering yields of aluminium bombarded with a range of ions and energies. It should be noted that these are not monotonic functions of mass and energy as the curves cross over and show maxima.

Because there are uncertainties in choosing the surface binding energy, U_0, it may be appropriate to consider how successful the theory is in predicting relative sputtering yield for a single target as a function of incident ions (Z_1). This is not without problems (e.g. Andersen and Bay 1972a, 1973 and Andersen 1973 and Figs 6.3 and 6.4) if the impinging ion alters the sputtering yield. At fluxes and energies where these problems are minimised the Z_1 dependence of the sputtering yield, relative to the self-sputtering yield ($M_1 = M_2$), is shown in Fig. 6.10 for targets of copper and silver. The experimental data is that of Andersen and Bay (1972b). Even if one compares data from a variety of sources there is general agreement with the collision cascade theory (Fig. 6.11).

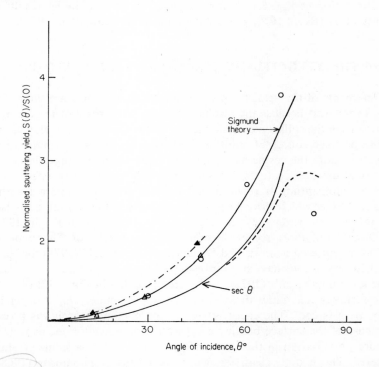

FIG. 6.12. The variation of sputtering yield with angle of ion incidence. The Sigmund (1969) prediction and sec θ curves are compared with experimental data of Cheney et al. (1965), (○○○); Colombie (1964), (△△△); Dupp et al. (1966), (▲▲▲); Molchanov et al. (1961), (– – –); and Rol et al. (1960), (– · – · –) for argon ions incident on copper.

6.5 THE ANGULAR DEPENDENCE OF THE SPUTTERING YIELD

The Sigmund (1969) theory also predicts a change in sputtering yield as a function of angle of incidence of the beam to the target. The yield rises at incidence angles away from the normal because there is a greater chance that the cascade is close to the surface. At very high angles of incidence the ion beam will merely bounce off the surface so that less energy is deposited and the yield decreases. The maximum in this sputtering yield typically occurs at an angle between 50 and 80° from the normal. The rate of change with angle (θ) is a function of M_2/M_1 and if $M_2 \gg M_1$ the dependence is roughly $(\cos \theta)^{-1}$ as was predicted by Thompson's (1968) model. However if the masses are more nearly equal the angular dependence changes towards $(\cos \theta)^{-5/3}$. A variation of sputtering yield with angle is shown in Fig. 6.12 for data from a variety of sources. This behaviour seems to be quite general for all isotropic targets, ions and energies which have been reported so far. The illustration of Fig. 6.12 is for a metal, copper, but similar results exist for insulators (Bach, 1970, Edwin, 1973 and Kanekama et al. 1973).

6.6 THE VELOCITY SPECTRUM OF EJECTED ATOMS

Measurements of the velocity spectrum of sputtered atoms are only of relevance to this text because the results provide a further test of the theoretical predictions of the collision cascade theory. They do indeed show that cascades are the primary source of sputtered particles. In crystallographic material one detects both the cosine-like angular distribution and a superposed set of directional ejection features. These correspond to focussing effects of the lattice as momentum or replacement collision sequences interact with the surface. Atoms which have been channelled from the crystal will also be seen in such measurements. A review of the topic was made in 1970 by MacDonald but more recent data includes the results of Chapman et al. (1972) for sputtered neutrals and Dennis and MacDonald (1972) for charged ions. Clusters of ions have been investigated by Standenmaier (1972). Some of the Chapman et al.'s (1972) results are presented in Fig. 6.13(a). Beyond the low energy maximum these follow an approximately E_0^{-2} dependence if the target is cool. The energy at which the maximum occurs is probably a measure of the surface binding energy and consequently one expects, and measures, a variation in the energy if different facets of a single crystal are sputtered. This is quite clearly shown in the results of Thompson (1968) for a gold single crystal bombarded with argon ions, Fig. 6.13(b). This figure presents the data in terms of both the time of flight spectra and the ion energy. The number of ions at a particular energy are described by a linear scale on Fig. 6.13(b) and a logarithmic scale on Fig. 6.13(a). To some extent this obscures the fact that on sputtering at high temperatures (Fig. 6.13(a)) not only does the total sputtering yield increase markedly but also the energy spectrum of the sputtered ions is peaked at much lower energies. The major fraction of this increase is in the random component of the sputtering although there is some broadening of the crystallographic features. The enhancement of sputtering rate at high temperatures may well result from a contribution from thermal spikes which were unimportant in low temperature experiments. Additional features in the spectra have been attributed to the effects of correlated collision, focussing and channelling. Such experiments are difficult and there are still many unresolved problems. Surface scattering events also change the energy spectrum (Reid *et al.*, 1976).

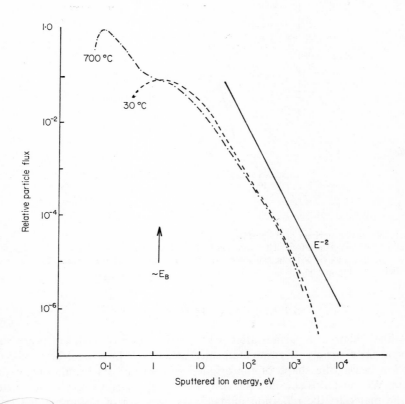

FIG. 6.13a. The energy spectra of gold atoms produced by bombardment with 20 keV argon ions. The sample temperatures were 30 and 700°C. For comparison the estimated surface binding energy E_B and theoretical E^{-2} distribution of particle energies are also shown.

Not the least of these is the problem of controlling the surface topography since this can completely alter crystallographic, directional and energy spectra and also change the relative fractions of neutrals to charged atoms and the ratio of single to multiple atom clusters in a single sputtering event.

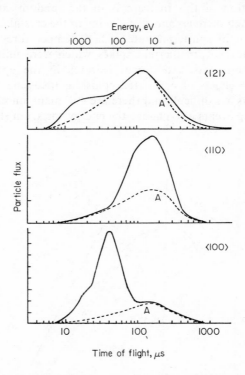

FIG. 6.13b. A comparison of time of flight spectra for atoms sputtered from single crystals of gold by 43 keV argon ions. The curves labelled A are predictions from a random cascade theory. Thompson (1968) suggested the other contributions might include focussing or assisted focussing events.

6.7 SURFACE TOPOGRAPHY

The final shape of a surface after it has been etched with an ion beam will depend on a variety of parameters such as surface diffusion; the profile of the ion beam; the shape of the original surface and its inclination to the beam. We must also consider the cleanliness of the surface; the homogeneity of the material; the radiation damage; the position of the implanted ions and the total amount of material which is sputtered. In the previous discussions of sputtering yields from plane isotropic solids we noted that the sputtering yield was a function of M_1, M_2, N and θ. Consequently, imperfections in the isotropic material will be revealed because they will sputter at a different rate. Also, if the material is isotropic but not planar then we will produce a new topography after sputtering.

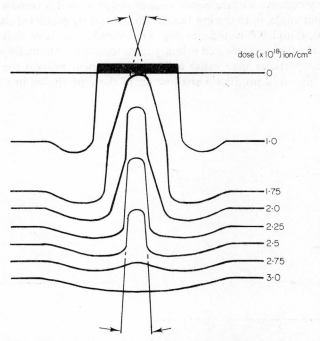

FIG. 6.14. The profile of a surface of GaAs which was partially covered by a region of lower sputtering coefficient. The changes in surface feature with ion dose produce a cone shaped feature surrounded by a region of enhanced sputtering at the base.

As a first example consider surface contaminants which have a lower sputtering rate than the substrate and so act as a shield for the material beneath. This will produce cones or mesa like features which are directly analogous with the geological mesas. Figure 6.14 shows the development of a cone-like feature on a surface which was initially flat but was decorated with a particle of lower sputtering rate. This particular figure was taken from scanning electron microscope data of Wilson (1973) where he bombarded GaAs with 40 keV argon ions. Many similar examples exist (e.g. Stewart and Thompson, 1969; Wilson, 1974). The flat topped feature is preserved until the dirt has eroded and then the residual hill develops angled sides, these sputter at a different rate from the base plane if the $S(\theta)$ curve ($\approx S(0) \sec \theta$) is not compensated by the change in flux on the angled face ($\Phi \cos \theta$). Depending on the relative magnitudes of these terms the cones may either grow or shrink (e.g. Todorescu and Vasilu, 1972). Stewart and Thompson (1969) considered that because there is a maximum in the $S(\theta)$ curve (Fig. 6.15) any concave surface would change shape until it formed a plane at the maximum angle, $\hat{\theta}$, to the ion beam. They therefore predicted that the planes at $\theta = 0$, $\hat{\theta}$ and 90° would develop. One should also note that as the cone shrinks the forward reflected primary ions, together with the forward directed sputtered particles may cause a flux enhancement around the base of the cone. This could produce a shallow pit of the type shown in Fig. 6.14.

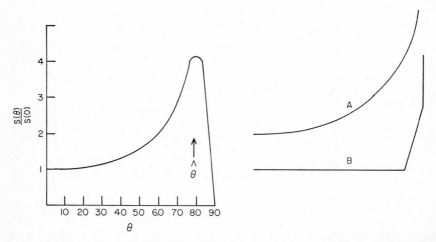

FIG. 6.15. The variation of sputtering yield with angle of incidence of the ion beam. Also shown is a concave surface (A) which sputters to the more stable planes (B) at $\theta = 0, \hat{\theta}, 90°$.

Cones of this type which result from non-uniform surfaces will always lie parallel to the ion beam in isotropic material. Impurities from within the solid may coagulate to form similar cones during erosion (Wehner and Hajicek, 1971). Stewart and Thompson (1969) considered the sideways motion of a plane which is not perpendicular to the ion beam. Figure 6.16 shows two planes A, B, at angles of incidence α and β to the ion beam. For the example shown the angle AOB is obtuse. These planes sputter to a new position A', B' and the line of intersection moves from 0 to 0'. This line forms an angle δ to the ion beam. If the incident flux is Φ ions per unit

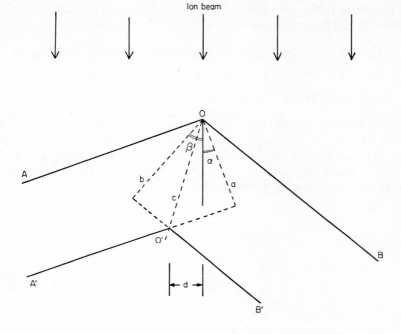

FIG. 6.16. The position of intersecting planes before and after sputtering.

area then each plane is bombarded with fluxes of $\Phi \cos \alpha$ or $\Phi \cos \beta$. So from the figure we see

$$a = \frac{\Phi \cos \alpha}{N} S(\alpha)$$

$$b = \frac{\Phi \cos \beta}{N} S(\beta) \quad (N \text{ is the atomic density})$$

Also

$$a = c \cos(\alpha + \delta)$$
$$b = c \cos(\beta - \delta)$$

and we can express

$$\frac{d}{a} = \frac{S(\beta) - S(\alpha)}{S(\alpha) \cos \alpha (\tan \alpha + \tan \beta)}.$$

This is useful because we can now decide whether the intersection moves to the left or right during sputtering. The possible cases are

(i) $S(\alpha) = S(\beta)$, $d = 0$ and $00'$ is parallel to the incident beam.

(ii) $S(\alpha) > S(\beta)$, $d < 0$ and $00'$ is a line moving to the right.

(iii) $S(\alpha) < S(\beta)$, $d > 0$ and $00'$ is a line moving to the left.

This predicts that the crest of a ridge will move towards the side for which $S(\theta)$ is less whereas the base of a valley will move towards the side for which $S(\theta)$ is greater. In the case of a surface step, Fig. 6.17, $\alpha = 0$ and β approaches $\hat{\theta}$. The structure will develop into a set of planes with the slope travelling across the sample during bombardment.

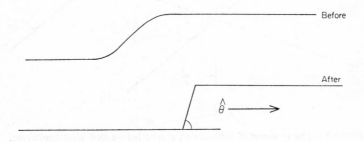

FIG. 6.17. The development of a surface step with planes at $\theta = 0$ and $\hat{\theta}$ after bombardment of a curved step. Note the plane at $\hat{\theta}$ will continue to cross the surface.

A more general analytical approach was attempted by Catana *et al.*, (1972) who showed the time dependence of the angle between the surface and the beam was analogous to a wave equation. They attempted to find solutions for the equation assuming that there was an initial sinusoidal character to the surface roughness. Unfortunately their choice of boundary conditions introduced artefacts in the final surface topography.

6.8 A COMPARISON OF SPUTTERING AND CHEMICAL DISSOLUTION

For isotropic solids there is a distinct similarity between chemical and ion beam etching. Barber *et al.* (1973) realised this and applied Frank's (1958) kinematic theory of dissolution to sputter etching. He considered the surface as a succession of small steps (of equal height) and termed dissolution, or sputtering in our case, as the rate at which these steps pass a reference point.

6.8 A COMPARISON OF SPUTTERING AND CHEMICAL DISSOLUTION

If k is the step density and q the number passing a fixed point then the slope of the surface

$$k = \left(\frac{\partial y}{\partial x}\right)_t \text{ and the sputtering rate is}$$

$$q = -\left(\frac{\partial y}{\partial t}\right)_x.$$

In one dimension, if no new steps are created or destroyed, $dk.dx = -dq.dt$. Also we can define a wave velocity

$$C(k) = \left(\frac{\partial q}{\partial k}\right)_x = \frac{dx}{dt}.$$

Combining these expressions gives

$$C(k)\left(\frac{\partial k}{\partial x}\right)_t + \left(\frac{\partial k}{\partial t}\right)_x = 0.$$

This means that in an (x, t) plane if a slope k remains constant then so does the flux q.

If we now consider a surface which changes with time, $y = y(x, t)$, then

$$\frac{dy}{dx} = \left(\frac{\partial y}{\partial x}\right)_t + \left(\frac{\partial y}{\partial t}\right)_x \frac{dt}{dx}$$

$$= k - \frac{q}{C(k)}$$

which we interpret as meaning that an initial slope proceeds along a fixed trajectory during sputtering.

Discontinuities are incorporated in the theory by specifying planes adjacent to the edge. The velocity is

$$\frac{dx}{dt} = \frac{q_2 - q_1}{k_2 - k_1} \text{ and the trajectory is}$$

$$\frac{dy}{dx} = k_2 - q_2 \left(\frac{k_2 - k_1}{q_2 - q_1}\right).$$

Frank's (1958) theory can be summarised by the following two theorems.

(i) *The locus of an elemental area of a crystal surface with a particular orientation is a straight line during etching (assuming that the etch rate is only a function of orientation). This line is termed a dissolution trajectory.*

(ii) *The trajectory of this elemental area is parallel to the normal of the polar diagram of the reciprocal of the etch rate at the point of similar orientation (defining the etch rate as being measured normal to the crystal surface).*

A corollary to these theorems is that we can compute the direction and dissolution rate of a surface discontinuity (e.g. an edge) by constructing the chord in the polar diagram which joins the points corresponding to the orientation of the planes adjacent to the discontinuity. The dissolution trajectory is parallel to the normal of this chord.

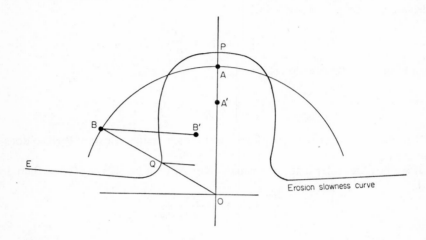

FIG. 6.18. An example of the use of the polar plot of the erosion slowness curve, E, to calculate the movement of two points on a sphere (A, B) to new positions A', B'. The displacements are proportional to $(1/OP)$ and $(1/OQ)$ and in direction parallel to the normal of the erosion slowness curve at the points P and Q.

Barber *et al.* (1973) deduce the changes in surface topography during sputtering by a geometric construction. We allow for the changes in flux and sputtering rate at each plane by defining a function $S(0)/(S(\theta) \cos \theta)$. Where $S(0)$ is the sputtering rate at normal incidence. This inverse function is called the "erosion-slowness" curve. We may now follow the changes in surface structure by the construction of Fig. 6.18 in which we compare the surface contour and the polar diagram of the erosion-slowness curve. Consider a point A on the surface. This area erodes at a rate proportional to $(1/OP)$ and in a direction parallel to the normal of the erosion slowness curve at the point P. Hence A will move to A' whereas B will move to B'.

For more detailed examples we commence with the erosion of a sphere, Fig. 6.19. The Barber *et al.* (1973) construction generates cone shapes as the

6.8 A COMPARISON OF SPUTTERING AND CHEMICAL DISSOLUTION

sphere sputters. For clarity we have presented both the details of the construction and a sketch of the original and final surfaces. This structure is in general agreement with the simple predictions of the Stewart and Thompson (1969) theory but we can now quantitatively analyse the development of the more complex structures.

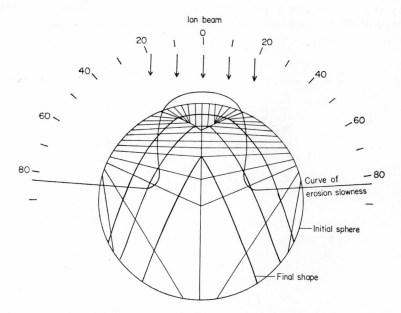

FIG. 6.19. The geometrical construction which indicates that the erosion slowness curve applied to a sphere will result in a cone shaped structure.

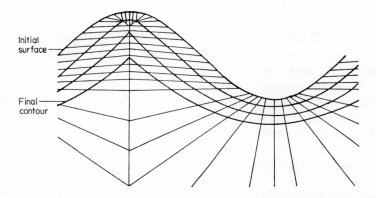

FIG. 6.20. The development of surfaces by sputtering a sinusoidal surface.

136 SPUTTERING 6.9

The sinusoidal surface, Fig. 6.20, also develops cones, beneath the maxima but the cone height is reduced with time. Less obvious is the fact that a circular indentation on the surface will be removed by ion beam sputtering, only if the trough is less than a complete semicircle. Figure 6.21 shows that a small pit will expand and flatten into the plane of the background whereas the complete semicircular pit will develop into a square pit. Once this stage is reached the top and base planes will move at identical speeds.

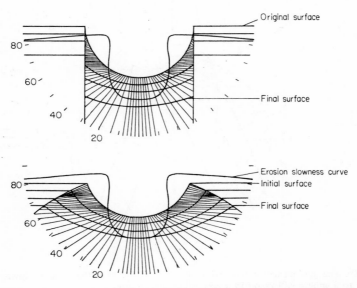

FIG. 6.21. The trajectories, erosion slowness curve, initial and later stages of (a) a full and (b) a partial hemispherical trough which is sputtered.

Simple bombardment of a surface with ions from a single direction cannot remove all surface features. Even rotation of the specimen may not remove all hillocks although the ratio of height to diameter will decrease with erosion time. For practical purposes of surface polishing changes in angle of incidence and specimen rotation will remove the majority of the surface features. One could of course use very fine ion beams to selectively sputter a limited region of the surface.

6.9 SECONDARY PROCESSES IN SURFACE TOPOGRAPHY

The preceding theory adequately describes the macroscopic changes in the surface topography but since it assumes an isotropic solid and ignores the details of the collision cascade or the fate of the ejected atoms it can not predict some of the smaller scale features nor the development of features related to secondary processes.

Sigmund (1973) pointed out that if the ion beam is not normal to the surface the collision cascade will produce a maximum sputtering yield further down the slope from the point of impact of the ion. This is clearly shown in Fig. 6.22 where the contour map of the radiation damage (i.e. the energy deposited with depth) intersects the surface near the maximum of the collision cascade. The point C on the figure will therefore sputter more rapidly than the point 0. Such an effect could generate surface roughness on the scale of the collision cascade (say 100 Å) which will not be removed by ion beam polishing. However, a further consideration might be to ask if surface migration could smooth out these small features. Certainly the *in situ* measurements of metal foils bombarded by ions in an electron microscope (Nelson and Mazey, 1973) show that even large surface craters, formed as a result of bubbles bursting at the surface, are rapidly smoothed over by atomic diffusion. Therefore this fine microstructure may not be of practical consequence. It should also be noted that with any conventional machining techniques we would be worried by much more gross features than the 100 Å scale of this fine structure.

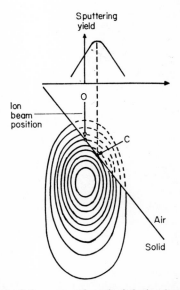

FIG. 6.22. A contour map of the energy deposited during ion bombardment and the consequent sputtering yield. Note the maximum in the yield curve is further down the slope than the point at which the ion was incident.

Sigmund's (1974) predictions of fine scale roughness are minor limitations to the ion beam applications of machining or polishing. The secondary effects of (i) flux enhancement produced by forward directed sputtered atoms plus reflected energetic primaries and (ii) the redeposition of sputtered atoms are more serious. Bayly (1972) considered both processes and presented results from the sputtering of silica to demonstrate their existence.

Bayly (1972) used silica specimens which had been mechanically polished, this leaves regions of mechanical strain which can develop as surface cracks

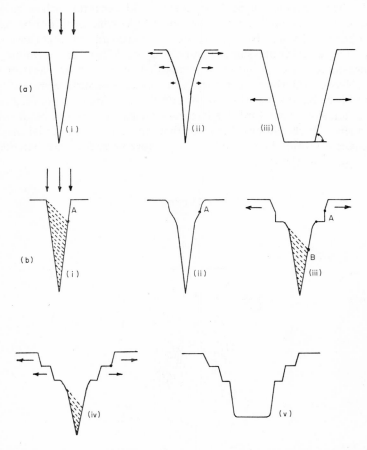

FIG. 6.23. Changes in the shape of a deep surface flaw with sputtering. (a) (i) the original feature; (ii) convex sides develop; (iii) equilibrium planes. (b) (i) the secondary flux enhances the erosion rate at the point A; (ii) an inflection point develops near A; (iii) and (iv) surface terraces develop. Also at point B redeposition tends to fill the base of the crack. (v) a final structure.

called Griffith's flaws. These latent cracks can be made visible by chemical etching in HF. Subsequent ion beam polishing produced a widening of the narrow fissures into pits with a flat base and terraced walls, as sketched in Fig. 6.23. Typical dimensions are widths of 4 microns and a depth of 1 micron.

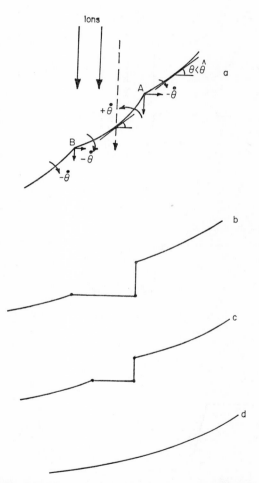

FIG. 6.24. The formation of a step from a secondary pit in a curved surface. Note that the angle of the slope to the incident beam is less than $\hat{\theta}$ for the major surface. Curves b, c, d show how the pit develops and finally shrinks. Points near A and B move with different senses of rotation according to the arrows for $\dot{\theta}$.

From the work of Stewart and Thompson (1969) it was expected that the sideways velocity of a plane at an angle θ to the incident beam would be

$$V = \frac{\Phi}{N}(S(\theta) - S(0))\cot\theta$$

and if the surface were curved the rotation of the plane at each point would

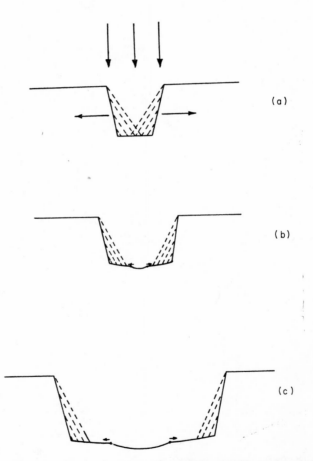

FIG. 6.25. Flux enhancement at the bottom of a steep walled pit because of forward reflected primary ions and forward ejected sputtered ions. Sketches (a), (b), (c) show the double enhancement of the bombardment from both walls in a narrow pit. As the walls recede sideways the flux reduces to the normal ion flux.

be given by (e.g. Carter et al., 1971)

$$\left.\frac{\partial \theta}{\partial t}\right)_x = -\frac{\Phi}{N}\cos^2\theta \frac{\partial S}{\partial \theta}\left.\frac{\partial \theta}{\partial x}\right)_t.$$

From Fig. 6.24 it is clear that even if a concave surface had a secondary pit within it such a feature should disappear during prolonged sputtering, so could not produce the observed terraces. (The figure also indicates the sense of rotation for the various parts of the surface.) Bayly (1972) therefore considered the flux enhancement at the bottom of a narrow pit, Fig. 6.25, where the primary flux is increased by forward scattered ions. Initially the narrow fissure is enhanced from both walls and so forms a central depression.

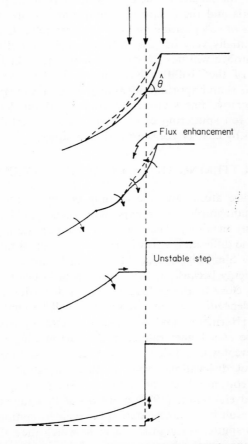

FIG. 6.26. Flux enhancement on a curved surface which leads to a secondary pit and hence a step.

However, as the walls recede sideways the enhancement is reduced. In a more normal case of curved side walls, Fig. 6.26 the enhanced flux develops a secondary pit which becomes a step. A large surface may therefore develop a series of terraces with flat steps and steep sides. Which is just the result experimentally observed. Large shallow pits will also develop a stepped rim because at the rim ions reflected from planes with $\theta \gg \theta_{max}$ strike a region further down the slope and produce enhanced erosion but the rotation of the planes is to the higher values of θ so the step is confined to the rim of the crater.

If the fissure is very narrow then the sputtered material may be redeposited on the opposite surface. Many atoms from the collision cascade are energetic enough to bond onto the surface so the narrowest parts of the pit (Fig. 6.23) will collect this debris and the fissure will disappear. Figure 6.23 explains how these two effects of redeposition and terracing are related.

If the fissure is initially very narrow then redeposition will dominate so that the sputtered surface will not reveal the latent crack. Bayly (1972) did not observe traces of the Griffith's flaws unless he had first chemically developed them. This is an important advantage of ion beam sputtering that it can be used to remove fine scale strain cracks without developing any additional features. Ion sputtering can be used as a powerful final polish for materials shaped by coarser conventional techniques.

6.10 SPUTTERING OF CRYSTALLINE SOLIDS

The density of surface atoms and the binding energy to the surface will depend on the crystallography and composition of the solid (i.e. the $S(\theta)$ curve will have many maxima). Therefore ion beam sputtering of crystals will develop facets and different grains will sputter at different rates according to their orientation. Similarly one could treat dislocations as regions of different bonding energy because of the mismatch of lattice planes and the distortion of planes close to the dislocation. Such distortion may increase the rate of energy deposition because fewer ions will be channelled along the strained planes (Hermanne and Art, 1970). Sputtering from the dislocation region will be at a higher rate than the surrounding planes so ion beam etching will reveal where dislocations intersect the surface. Even if the surface were without dislocations prior to the irradiation they may be generated by high concentrations of radiation damage (e.g. platelets of interstitials, etc.) and Hermanne (1973) pointed out that radiation induced dislocations would result if the time required for their formation is less than the time required to sputter through the layer in which they are produced. For the same reasons phase changes produced by implantation will be detectable by variations in the sputtering yield.

In some materials impurities segregate to the dislocation lines. This produces a more complex sputtering situation with a pit developing around the dislocation, crystal facets determine the orientation of the walls of the pit but the impurity ion which previously decorated the dislocation line now aggregate into a cluster of atoms. This is the classical "dirt" situation of Stewart and Thompson (1969) and so leads to cone formation in the centre of the pit. The overall picture, sketched in Fig. 6.27, has been observed in the electron sputtering experiments of alkali halides by Elliott and Townsend (1971).

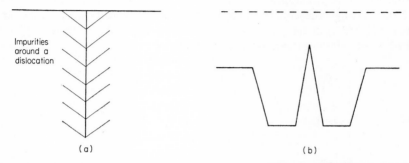

FIG. 6.27. Sputtering of a region which contains an impurity decorated dislocation. (b) shows how the reduced sputtering coefficient of the impurities leads to cone formation. Whilst the surrounds sputter more rapidly because of flux enhancement and the higher sputtering rate of the strained lattice.

6.11 TRANSMISSION SPUTTERING

Ion ranges are rarely more than a few microns in length so sputtering from the rear face of a bombarded specimen will only occur if the sample is thin. Channelling or the presence of cones changes this situation by a small amount. However, if a thin film is deposited on a substrate the collision cascade can transfer momentum, in the form of moving atoms, across the boundary. This mixing of the layers can be assisted by radiation enhanced diffusion. Transmission sputtering in this situation can provide adhesion between the two layers.

6.12 SPUTTER DEPOSITION

Our attention so far has been on the sputtering yield and the surface topography and we have barely mentioned the energy spectrum of the ejected atoms or the ejection patterns from crystalline targets. However, the process of sputtering provides us with a unique source of moderately energetic atoms. The energies of tens of eV may be low compared with those considered in implantation but such energies are unobtainable from thermal evaporation of atoms. Sputtering can thus provide the source material for coatings or plating applications. There are several advantages in choosing this type of deposition source and they include:

(i) The greater energy of atoms from a sputtering source, compared with a thermal source, produces a simultaneous cleaning and deposition of the target surface. A comparison of the velocity distributions of atoms from the two types of source are shown in Fig. 6.28. Enhanced cleanliness and higher energies results in a surface with good adhesion properties and reproducible characteristics (e.g. Chopra, 1969). Also the atoms are sufficiently energetic that they change the sticking probability and concentration of nucleation sites compared with thermal evaporation. This means single crystal epitaxy can occur at different substrate temperatures.

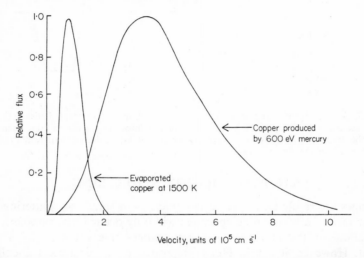

FIG. 6.28. A comparison of the velocity spectra of copper atoms produced by evaporation (at 1500 K) and sputtering (with 600 eV Hg^+ ions).

(ii) It is possible to sputter alloys and compounds, both insulating and conducting. The composition of the deposited layer can be altered by simultaneous sputtering of two sources or modified by the presence of a background gas. Cosputtering rates are more readily controlled than thermal evaporation from two independent sources so the resultant layer may be uniform and prepared in a chosen phase.

(iii) It is chemically cleaner than many coating or plating processes and diffusion can be encouraged or suppressed by control of the substrate temperature.

The advantages of sputtering sources for the deposition of thin films have been used for many years and greatly outweigh the minor problems. One such problem is that in sputtering an alloy one component may sputter preferentially and so there will be an induction time, set by the relative emission rates and diffusion coefficients, before the sputtered particles match the stoichiometry of the source. A second problem exists if the source is small and the target is contoured so that there is shadowing of the target. This can be avoided by rotating the target or increasing the background gas pressure to produce multiple collisions and hence an effectively more random source. There are clearly optimum conditions of gas pressure, ion current and energy for high deposition rates. A detailed prediction is difficult since the multiple collisions with the background gas alter the energy distribution of the atoms and hence their sticking probability on the target. Many detailed examples of sputter deposition have been quoted, together with the operating conditions, in the many review articles on this subject (e.g. Berry et al., 1968; Chambers and Carmichael, 1971; Chopra, 1969; Holland, 1965, 1970, 1972; Holland et al., 1974; Kennedy et al., 1971; Wan et al., 1971, also the Proceedings of Symposia on Deposition of Thin Films by Sputtering).

In multicomponent deposition the sputtering method allows one to codeposit normally incompatible systems such as metal–non–metal mixtures or non-equilibrium alloys. The surface mobility of the deposit can be reduced by cooling of the target. This may stabilise the mixed deposit.

Chemical reactions with the background gas also control the composition of the sputtered deposit. Changes can occur during the passage of the sputtered atoms through the gas or by chemical reactions between the gas and the newly deposited excited surface. This adds additional flexibility in the choice of the sputtered layer.

Compound sputtering can have problems of non-stoichiometry for example SiO_2 may appear as a sputter deposit in the form $SiO_{1.6}$. To offset this oxygen deficiency one can carry out the sputtering in a gas mixture containing oxygen. Chemical reactions of this form are termed reactive sputtering and have been used in the production of oxides, nitrides, carbides and sulphides. With such insulators the gas pressure during sputtering also allows control of the electrical conductivity of the layer.

Because reactions between the gas and sputtered atoms can lead to ion formation the deposition in the presence of a gas has also been termed ion plating (Mattox, 1964).

6.13 CONCLUSION

The underlying processes in sputtering are well understood and despite the complexity of the problem quantitative predictions of the sputtering yield can be made. If the process is to be used for machining or polishing application then one must also consider the changes in topography which can occur. The technological applications of sputtering and sputter coating are both numerous and diverse as will be demonstrated in Chapter 9 where we will mention a selection of applications ranging from the polishing of aspheric lenses, to hole drilling, to the detection of diseased blood cells.

REFERENCES

Almén, O. and Bruce, G. (1961). *Nucl. Inst. and Meth.* **11,** 257 and 279.
Andersen, H. H. (1971). *Rad. Effects*, **7,** 179.
Andersen, H. H. (1973). see Behrisch *et al.* p. 71.
Andersen, H. H. and Bay, H. L. (1972a). *Rad. Effects*, **13,** 67.
Andersen, H. H. and Bay, H. L. (1972b). see Andersen *et al.* p. 313.
Andersen, H. H. and Bay, H. L. (1973). see Behrisch *et al.* p. 63.
Andersen, S., Björkqvist, K., Domeij, H. and Johansson, N. G. E. (Eds). (1972). "Atomic Collisions in Solids IV", Gordon and Breach. These papers are the proceedings of a conference at Gausdal and also appear in Radiation Effects.
Bach, H. (1970). *J. Non-Cryst-Solids*, **3,** 1.
Barber, D. J., Frank, F. C., Moss, M., Steeds, J. W. and Tsong, I. S. T. (1973). *J. Mat. Sci.* **8,** 1030.
Bayly, A. R. (1972). *J. Mat. Sci.* **7,** 404.
Behrisch, R., Heiland, W., Poschenrieder, W., Staib, P. and Verbeck, H. (Eds). "Ion Surface Interaction, Sputtering and Related Phenomena", Gordon and Breach. These papers also appeared in *Radiation Effects*, 1973, vols. 18 and 19.
Berry, R. W., Hall, P. M. and Harris, M. T. (1968). "Thin Film Technology", Van Nostrand.
Carter, G. and Colligon, J. S. (1968). "Ion Bombardment of Solids", Heinemann.
Carter, G., Colligon, J. S. and Nobes, M. J. (1971). *J. Mat. Sci.* **6,** 115.
Catana, C., Colligon, J. S. and Carter, G. (1972). *J. Mat. Sci.* **7,** 467.
Chambers, D. L. and Carmichael, D. C. (1971). *Res. and Development Mag.* 22.
Chapman, G. E., Farmery, B. W., Thompson, M. W. and Wilson, I. H. (1972). see Andersen *et al.*, p. 339.
Cheney, K. B. and Pitkin, E. T. (1965). *J. Appl. Phys.* **36,** 3542.
Chopra, K. L. (1969). "Thin Film Phenomena", McGraw-Hill.
Colombie, N. (1964). Thesis, University of Toulouse.
Dennis, E. and MacDonald, R. J. (1972). See Andersen *et al.* (1972), p. 333.
Dupp, G. and Scharmann, A. (1966). *Z. Phys.* **192,** 284 ibid **194,** 448.
Edwin, R. P. (1973). *J. Phys. D.* **6,** 833.
Elliott, D. J. and Townsend, P. D. (1971). *Phil. Mag.* **23,** 249.
Frank, F. C. (1958). "Growth and Perfection of Crystals" Doremus, R. H. Roberts, B. W. and Turnbull, D., (Eds). Wiley, p. 411.
Guseva, M. I. (1960). *Sov. Phys. Sol. State*, **1,** 1410.
Hermanne, N. (1973). *Rad. Effects*, **18,** 161.
Hermanne, N. and Art, A. (1970). *Rad. Effects*, **5,** 203.
Holland, L. (1965). "Thin Film Microelectronics", Chapman and Hall.
Holland, L. (1970). *Electronic Components*, March, 285; May 1.
Holland, L. (1972). *Electronic Components*, May, 193; August, 761.

REFERENCES

Holland, L., Steckelmacher, W. and Yarwood, J. (1974). "Vacuum Manual", Spon.
Honig, R. E. (1962). *R.C.A. Review*, **23**, 4.
Kaminsky, M. (1965). "Atomic and Ionic Impact Phenomena", Springer-Verlag.
Kanekama, N., Taniguchi, N., Watanabe, K., Kondo, M. and Matsumoto, T. (1973). *Sci. Papers. Inst. of Phys. and Chem. Research*, **67**, 25.
Kennedy, K. D., Scheuermann, G. R. and Smith, H. R. (1971). *Res. and Development Mag.* **22**, 40.
McCracken, G., (1975). Reports on Prog. in Phys. **38**, 241.
MacDonald, R. J. (1970). *Advances in Physics*, **19**, 457.
Mattox, D. M. (1964). *Electrochem. Technology*, **2**, 295.
Molchanov, V. A. and Tel'kouskii, V. G. (1961). *Sov. Phys. Doklady*, **6**, 137.
Navinsek, B. (1972). "VI Yugoslav Symp. on the Physics of Ionized Gases", p. 221.
Nelson, R. S. and Mazey, D. J. (1973). See Behrisch *et al.*, p. 199.
Norgate, P. and Hammond, V. J. (1974). *Phys. in Technology*, **5**, 186.
"Proceedings of Symposia on Deposition of Thin Films by Sputtering," Bendix Corp.
Reid, I., Farmery, B. W. and Thompson, M. W. (1976) *Proc. Surf. Sci.* June issue.
Rol, P. K., Fluit, J. M. and Kistemaker, J. (1960). *Physica*, **26**, 100.
Sigmund, P. (1969). *Phys. Rev.* **184**, 383.
Sigmund, P. (1972a). *Rev. Roum. Phys.* **17**, 823, 969, 1079.
Sigmund, P. (1972b). "VI Yugoslav Symp. on the Physics of Ionized Gases," 137.
Sigmund, P. (1973). *J. Mat. Sci.* **8**, 1545.
Sigmund, P., Matthies, M. T. and Phillips, D. L. (1971). *Rad. Effects*, **11**, 34.
Southern, A. L., Willis, W. R. and Robinson, M. T. (1963). *J. Appl. Phys.* **34**, 153.
Staudenmaier, G. (1972). See Andersen *et al.*, p. 353.
Stewart, A. D. G. and Thompson, M. W. (1969). *J. Mat. Sci.* **4**, 56.
Thompson, M. W. (1968). *Phil. Mag.* **18**, 377.
Todorescu, I. A. and Vasilu, F. (1972). *Rad. Effects*, **15**, 101.
Wan, C. T., Chambers, D. L. and Carmichael, D. C. (1971). *J. Vac. Sci. and Tech.* **8**, 99.
Wehner, G. K. (1957). *Phys. Rev.* **108**, 35.
Wehner, G. K., Stuart, R. V. and Rosenberg, D. (1961). General Annual Report of Sputtering Yields, No. 2243.
Wehner, G. K. and Hajicek, D. J. (1971). *J. Appl. Phys.* **42**, 1145.
Weijsenfeld, C. H. (1967). Philips Res. Reports Suppl. No. 2.
Wilson, I. H. (1973). *Rad. Effects*, **18**, 95.
Wilson, I. H. (1974). "VII Yugoslav Symp. on the Physics of Ionized Gases."
Yonts, O. C., Normand, C. E. and Harrison, D. E. (1960). *J. Appl. Phys.* **31**, 447.

CHAPTER 7

Ion Beam Equipment

7.1 ION ACCELERATORS

Particle accelerators have a long history and have developed from the few hundred volts used at the end of the last century by J. J. Thomson to accelerate electrons to the recent GeV machines used to probe the elementary particles. The accelerators used in ion implantation are nearer to Thomson's design than to the new machines of the nuclear physicist. The energy range of interest in implantation is less than a few hundred thousand electron volts and this voltage can easily be obtained in a number of ways by conventional electronics. Much higher energies in the MeV range are however needed for example, for Rutherford backscattering measurements of damage implant profile and for channelling studies.

Most previous particle accelerators with energies up to a few MeV were developed for nuclear physics and used protons, deuterons, helium or occasionally other gases, which were ionized in a radio frequency source and the ions accelerated down a flight path to which the voltage was applied. A cylindrical beam was invariably used and a number of lenses and quadrupoles are normally needed to bring a well focussed beam onto the target. It is unfortunate that many of the most useful ions we wish to implant are not gases and so more sophisticated ion sources have had to be developed. Many of the materials we wish to ionise are also corrosive, which further complicates the design of a reliable efficient source capable of generating milliamps of current from a wide range of ions for a useful length of time.

The manipulation of targets and the control of their temperature become a significant part of the design, as does the maintenance of a vacuum good enough to prevent surface contamination due to the cracking of oil vapour by the beam, or the deposition of other impurities, on the target surface.

The technical problem of placing the ion of our choice at the desired depth in our target and doing so uniformly over the target area thus resolves itself into four parts: (i) the design of the ion source, (ii) the acceleration and shaping of the ion beam, (iii) the separation of the chosen ion from other ions and atoms and (iv) the target chamber with beam scanning and target manipulation facilities. The successful ion implanting machine combines the above four elements in a matched and balanced manner and a variety of such configurations exist.

Beam steering and mass analysis is almost invariably magnetic and magnets of considerable size are involved, particularly for energetic heavy ion machines with a vertical accelerating column which require a 90° deflection of the beam. In horizontal machines a smaller deflection can be used and hence a smaller magnet. The magnet can also serve as a switch magnet to deflect the beam down a number of different beam lines.

The size of the magnet can be further reduced by analysing the beam before it is fully accelerated and only accelerating the chosen ion. This can considerably reduce the current load on the accelerator column and hence on the high tension supply at the expense of placing the analyser, or at least the beam line passing through it, at the high tension voltage. At the voltages with which we are concerned this is no real problem, control of circuits in the high tension terminal being easily achieved by insulated rods or optical links. Thus the beam entering the analyser does so at a fixed extraction voltage which simplifies the ion optics. The accelerator then copes with only the selected ion and this combination gives good performance over a wide energy range.

Another alternative is to apply a high tension voltage to the target. This approach has disadvantages if we are doing measurement on the particles scattered from the target as all the rather delicate detectors and their electronics must be floated up at the same voltage, but where we are merely firing ions into a target at a certain energy the floating potential on the chamber is useful. For example, the machine and analyser are undisturbed which avoids changes in focus and beam current and the depth profile can be programmed by changing the voltage on the target chamber. It offers the opportunity of extending the accelerator range above the high voltage limit of its power supply or below the voltage at which the accelerator optics focus well. The reader is referred to Dearnaley *et al.* (1973), or Wilson and Brewer (1973) for descriptions and numerous detailed drawings of existing machines in a variety of configurations. We will confine ourselves here to considering individually the more general aspects of the four component areas mentioned above and then summarise the way in which they are mated together to make an ion implantation machine for a specific purpose.

7.2 ION SOURCES

Any process which imparts to an atom energy greater than the first ionization potential can produce positive ions. The most common method of imparting the energy is by means of an electrical discharge in a gas of the chosen atoms. If we strip two electrons from the atom and produce a doubly charged positive ion we again achieve twice the energy for a given accelerating voltage. The yield however is considerably reduced and most machines operate with sources which produce predominantly singly charged positive ions.

To produce the currents required, a plasma discharge is usually needed and, as some decades of work on plasma physics has confirmed, a plasma is a difficult and unstable thing to understand and control. It is even more difficult to extract our ions from the plasma without upsetting it and in fact the characteristics of many sources have proved very susceptible to the extraction voltages and geometry. One cannot calculate completely the behaviour of such sources from first principles and the development of new sources is frequently semi-empirical in approach.

Most types of source use a magnetic field to concentrate the plasma or arc in the region of the extraction aperture. This field also has the advantage of coiling up the electron paths and, as in the Penning vacuum gauge, enabling the source to run at lower pressures. Any reduction in source pressure is an advantage as neutrals escaping from the source constitute a vacuum leak and increase the pumping requirements. A source of low ionizing efficiency hence requires large pumps near the source, and perhaps at the high voltage terminal, to maintain an adequate vacuum and minimum gas scattering in the adjacent beam line. Many ion sources are developments of gas ion sources and use atoms which are derived by vapourising a solid in an oven, by sputtering, or from compounds with high vapour pressure which dissociate in the source to provide the requisite free ion.

Most atoms form a volatile chloride and this is the basis of the widely used carbon tetrachloride methods, Sidenius and Skilbreid (1961) in which carbon tetrachloride is passed over the heated oxide of the element, making the volatile chloride *in situ*. The breakdown of the CCl_4 in the ion source can give a complex mass spectrum of fragments and it may be preferable to use chlorides themselves as starting material where they are available, as is the case with phosphorous, boron and tin for example. Useful summaries of appropriate source materials for most of the elements are given by Dearnaley *et al.* (1973), Wilson and Brewer (1973), Freeman and Sidenius (1973) together with the hazards and problems encountered in each case.

Negative ion sources accept the appropriate atoms, attach one or more electrons and emit negative ions. They are of great importance in tandem accelerators where both ion source and target are at earth and the terminal with its electron stripper is at the positive high voltage. Thus the ion receives twice the full high tension voltage as it is accelerated both as approaching the terminal as a negative ion and leaving it as a positive ion. Negative ions are more difficult to make in most cases and, at our energy range, the stripping technique is not a useful method. We have therefore confined ourselves to positive ion sources. Running time before routine servicing of all ion sources is at best a few hundred hours because apertures sputter and insulators become coated with conducting layers. Solid source materials may be replenished in many source designs through a vacuum lock (e.g. see Fig. 7.11).

7.3 RADIOFREQUENCY ION SOURCES

The most commonly used source in ion accelerators has been the one in which the gas at a pressure of 10^{-2}–10^{-3} Torr is excited to ionization by

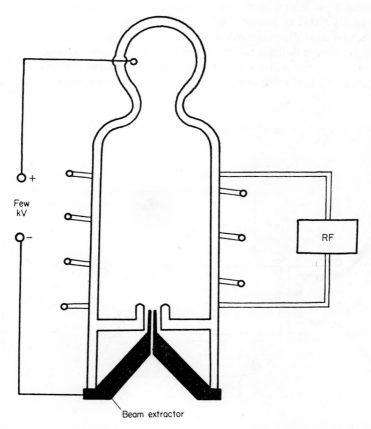

FIG. 7.1. The basic radiofrequency ion source. The gas in the glass tube is ionized by a radiofrequency supply and the positive ions move towards the beam extractor under the applied field. Not shown are the external magnets normally used to enforce and direct the discharge.

collisions produced by a few hundred watt radiofrequency supply of 10 to 100 MHz. Most are variations of the Thonemann *et al.* (1946, 1948) source, Fig. 7.1, which is essentially a glass tube surrounded by a radiofrequency coil and with an extractor like a Langmuir probe (Smith and Langmuir, 1926). The source works well for gases and can be used less successfully for solids by any of the vaporisation techniques mentioned above. More modern versions use capacitance rather than inductive coupling and a magnetic field to concentrate the discharge about the extractor. A few kilovolts positive potential are applied to the anode and the resultant field moves the positive ions towards the extraction aperture in the cathode.

Used with volatile salts of metals the inside of the tube becomes coated by a metal layer which limits the run time. The same applies to the use of a sputtered electrode to provide the metal ion, although simple easily replaceable glass tubes with sputterable metallic end caps have been used with some success on production machines. The run time has been extended to up to 100 hours with mA currents by placing an oven about the source Fig. 7.2 (Komrov *et al.* 1969) which vaporises a volatile salt of the ion required.

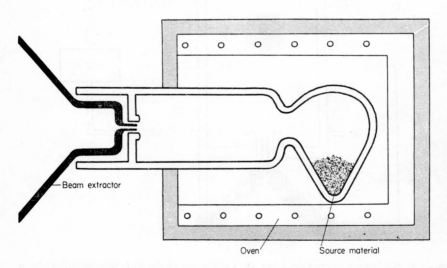

FIG. 7.2. Radiofrequency source for a volatile solid source material. An oven vaporises the source material and keeps the wall of the discharge tube hot enough to discourage redeposition. The radiofrequency exciting coils and magnets are not shown.

7.4 THE DUOPLASMATRON OR VON ARDENNE SOURCE

Von Ardenne's (1956) duoplasmatron achieves an efficiency of 50–95% by concentrating a magnetically confined arc through a small aperture in an intermediate electrode which is at a potential intermediate between that of anode and cathode, Fig. 7.3. The jet of plasma which emerges from the anode extraction aperture has a shape which depends critically on the operating conditions and geometry of the source and extraction electrodes, for example the quality of the subsequent image depends on whether its surface is concave or convex, Fig. 7.4. A number of commercial sources are available and an extensive literature exists on the use of the duoplasmatron with gases (Wilson and Brewer, 1973). From our point of view its main feature is an intensely ionizing environment with an ion density of 10^{14} cm^{-3} in the concentrated arc region which is capable of ionizing anything fed into it. This has been used by Masic *et al.* (1969) who enclose the plasma cap in a heated cup, Fig. 7.5, into which the material to be ionised is

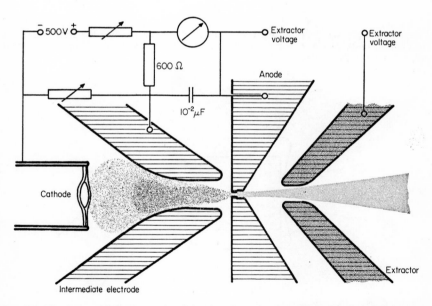

FIG. 7.3. The intermediate electrode of the duoplasmatron confines the discharge between cathode and anode to a small region and thus produces intense ionisation. The extractor accelerates ions from the plasma cap which projects through the anode aperture (after von Ardenne, 1956).

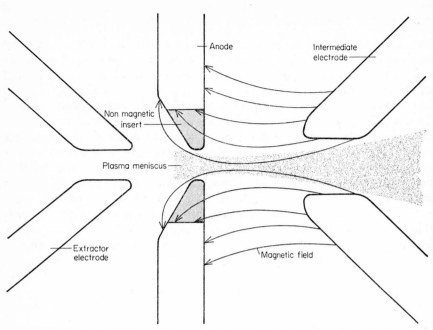

Fig. 7.4. The plasma is shaped by a magnetic field applied through the steel anode and steel intermediate electrode. The anode insert is normally tungsten or a similar high melting point material (after Brewer *et al.*, 1961).

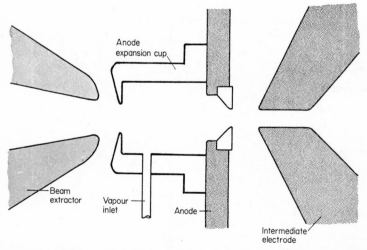

Fig. 7.5. Anode expansion cup fitted to a duoplasmatron in which some of the vapour introduced will be ionised by the plasma cap and extracted.

7.4 THE DUOPLASMATRON OR VON ARDENNE SOURCE

fed, with the advantage that for corrosive elements the main part of the source is free from attack. They report ion currents of up to 1 mA for Li, Cu and O_2. A disadvantage of this source when working with metal ions is that because of the comparatively large aperture of the plasma expansion cup, coating of insulators in the extraction region can lead to electrical breakdown. The vapour oven was hence moved to between the intermediate electrode and the anode by Illgen *et al.* (1972) and made a part of the intermediate electrode structure by Shimizu *et al.* (1973), Fig. 7.6. This boron nitride oven in the centre of the carbon intermediate electrode reduces the usual magnetic field produced by using the intermediate electrode and the anode as pole pieces, Fig. 7.4. With a pressure in the oven of 10^{-2}–10^{-3} Torr a consumption rate for silver was found to be 35 mg h^{-1} which gives a

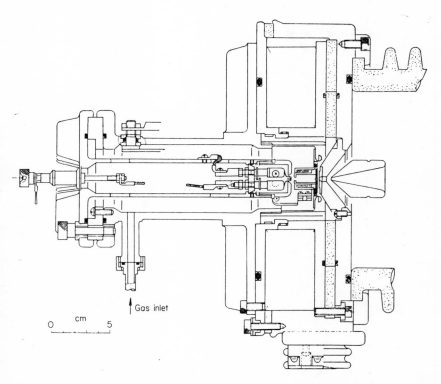

FIG. 7.6. Duoplasmatron with boron nitride vaporiser oven built into a carbon intermediate electrode, with metal source materials there is less trouble from insulator cooling and subsequent breakdowns in the extraction region for this geometry compared with that shown in Fig. 7.5 (after Shimizu *et al.*, 1973).

charge life of 100 hours. The run time however is limited by filament life to some 20 hours. Stable mA beams of silver, aluminium and copper have been obtained and the authors claim development to 10 mA is possible.

The power consumption of duoplasmatrons is of the order of a kilowatt, with a few hundred watts extra for sources with heated ovens. In spite of the disadvantage of this heavy power consumption it is a reliable well developed source and if it can now work well with a wide variety of ions, as seems likely from the recent developments mentioned above, it should find an increasing place in ion implantation machines, particularly if it can be developed in its ribbon beam form, von Ardenne (1961), to operate with heavy ions.

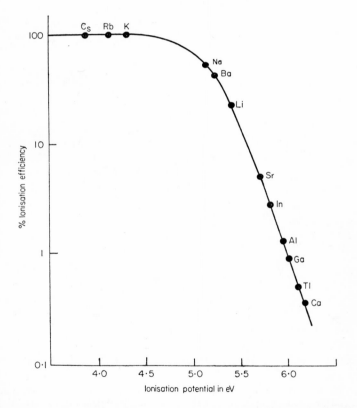

FIG. 7.7. The ionisation efficiency of a hot iridium surface ioniser at 2000 K for a number of elements.

7.5 SURFACE IONISATION SOURCES

A source which, unlike those mentioned above, does not rely on energetic collisions between atoms, ions and electrons is the surface ionisation source. The principle is simple. If an atom with a low ionisation potential is desorbed from a surface with a high work function it is more likely to lose an electron to the substrate and come off as a positive ion than as a neutral atom. The practice of producing alkali metal ions from heated tungsten surfaces has been used for decades in both ion sources and detectors for neutral alkali atoms. The ionisation efficiency, η, is given by the Saha–Langmuir (Langmuir and Taylor, 1933) equation,

$$\eta^{-1} = 1 + \omega \exp[e(I - \phi)/kT]$$

where ω is the statistical weight ratio of atoms to ions on the adsorbed substance, I is the ionization potential and ϕ is the substrate work function. A large $(I - \phi)$ hence means a high ionization efficiency. Thus a source using an indium substrate with an almost 100% efficiency for caesium will have less than a 0·5% efficiency for calcium, Fig. 7.7.

The ioniser surface must be heated to such a temperature that the desorption rate of the substance to be ionized is greater than its arrival rate, otherwise the surface will become covered by the atoms of the material we are attempting to ionise and the excess potential $(I - \phi)$ will disappear. At this critical temperature, at which arriving atoms are more likely to encounter a substrate atom than one of their own kind, a sharp increase in ion output occurs and then levels off. Except for a prebake to vaporise near surface impurities no advantage is to be gained from running an ioniser much above its critical temperature for a given rate of arrival of atoms. The need to maintain the ioniser surface at less than 100% coverage and the finite melting temperatures of the substrates imposes a fundamental limit on the ion current density attainable from a given substrate-adatom combination. Thus platinum with the high work function of 5·8 eV is limited as an ioniser by its melting point of 1769°C, whereas iridium with a work function of 5·3 eV and a melting point of 2443°C is commonly used. Higher work functions are available using non-metals but surface charging is a problem if the ioniser is not a good conductor and stability can be a problem on treated metallic surfaces such as oxygenated tungsten (Hague and Donaldson, 1963).

In spite of the limitations discussed above surface ionisation ion sources have specific advantages which make them useful. No support gas is needed, unlike sources which use an arc region for ionization. This gives much greater beam purity (particularly as they do not ionise the residual gas in the system) and greatly reduces the pumping problems. They are hence ideal for ultra high vacuum applications. The emerging beam also has a very low energy spread. The ionizer can be shaped to optimise the extraction ion optics and the shape is of course constant at all operating conditions unlike, for example, the duoplasmatron plasma cap. Sources of circular or strip configuration are readily made by choosing an ioniser of the appropriate shape. Strip beams with adequate uniformity have been produced by Wilson and Jamba (1967). Such a source is shown in Fig. 7.8, where the incident atomic beam from the oven impinges obliquely on the ioniser, the ion beam is steered out by appropriately shaped accelerating electrodes and the neutrals which escape ionisation are collected on a cold surface.

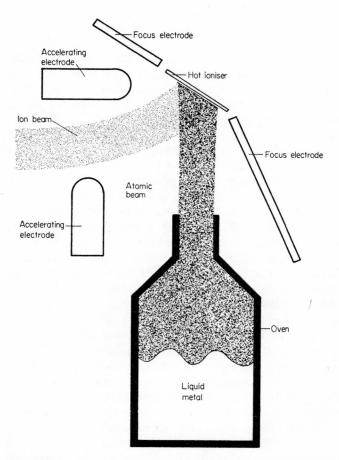

FIG. 7.8. Surface ionisation source using an atomic beam from an oven ionised by a hot surface ioniser. A variety of beam cross sections can be produced by suitably shaping the oven aperture and the hot ioniser.

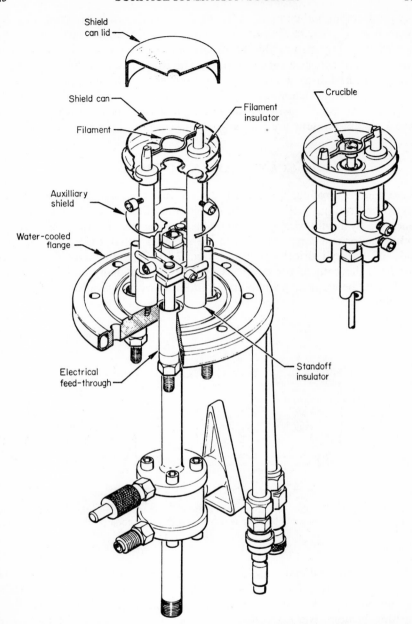

FIG. 7.9. A high temperature, up to 3300°C, source in which electron bombardment is used to vaporise the source material and ionise it at the hotter extraction orifice (after Johnson *et al.*, 1973).

A more recent and compact surface ionisation source is shown in Fig. 7.9, Johnson *et al.* (1973). The use of electron bombardment enables the exit hole of the tungsten crucible to be maintained at up to 3300°C at which temperature it acts as a surface ioniser. The base of the crucible is heated by conduction and is at a lower temperature, chosen to suit the material to be evaporated and determined by the crucible length.

Oxides are usually used as the source materials as they are stable and have a suitable vapour pressure at the temperature of operation.

7.6 SCANDINAVIAN SOURCES

These sources produce a conical beam and use an axial magnetic field and electron reflection, Nielsen (1957), Almén and Nielsen (1957). The electrons generated by a cylindrically wound filament are reflected between the cathode and extractor end caps which are at the same potential, the anode being the body of the cylindrical enclosure, Fig. 7.10. The magnetic field greatly lengthens the path of the electrons as they move radially and are reflected axially and hence their probability of colliding with a gas atom and ionizing it is enhanced.

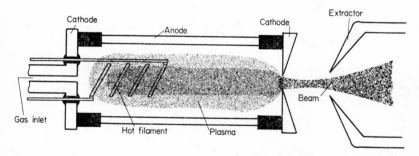

Fig. 7.10. A source in which electrons are confined by an axial magnetic field and reflected from the ends of the plasma chamber which are both at cathode potential (after Nielsen, 1957).

For use with solid materials, or the carbon tetrachloride method, a crucible is inserted axially into the source through a vacuum valve thus enabling replenishment of the charge, or a change of material without breaking the machine vacuum, Nielsen (1969). A number of developments of this source exist using different types of oven to vaporise the charge, or with modified electrode configurations, Lerner (1965), Magnuson *et al.* (1965).

A hollow cathode source devised by Sidenius (1969), Fig. 7.11 has the advantage of very compact size and operates up to a temperature of 2000°C which allows the direct vaporisation of many solid materials. The higher temperature elements are inserted into the source body and elements which vaporise more easily can be heated sufficiently by heat conducted along the gas feed tube in which they are placed.

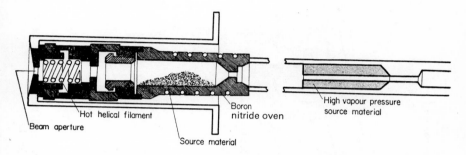

Fig. 7.11. Very compact hollow cathode source operating up to 2000°C and allowing direct vaporisation of many materials which are placed in the boron nitride oven. The helical filament is in the form of a coil spring with ends ground flat which is held in compression between its end contacts. The inhomogeneous magnetic field is not shown (after Sidenius, 1969).

An inhomogeneous magnetic field concentrates the plasma towards the extraction aperture and the discharge is controlled by a helical tungsten filament. Both types of source have been extensively developed and fitted to a number of machines throughout the world. They reliably produce currents of a few hundred microamperes of a wide range of elements.

7.7 HARWELL ION SOURCE

Another highly developed source is the Harwell heavy ion source, Freeman (1963), Fig. 7.12. It too has been fitted to many production ion implanters and is one of the most versatile ion implantation sources available. Unlike most of the sources already mentioned it produces a wedge shaped beam by using side rather than axial extraction from the anode cylindrical arc cavity. It produces milliamp beams rather than the tens or hundreds of microamps produced by most other sources. A magnetic field of about 100 gauss is applied parallel to the filament which, together with the field produced by the filament current spirals the emitted electrons around the filament giving a long path and hence greater ionisation. The long axial filament also stabilises the arc directly behind the extraction slit resulting in very stable operation

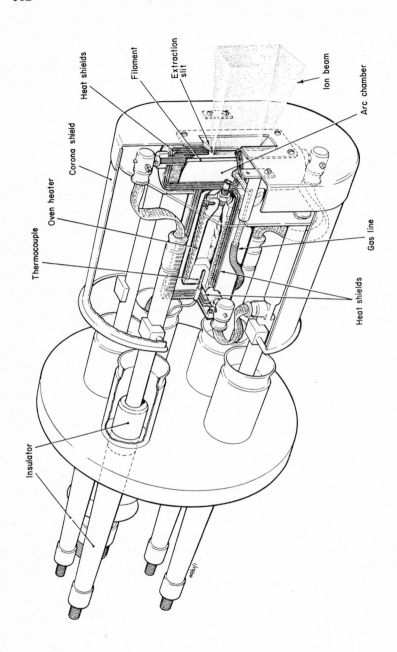

Fig. 7.12. Ion source with slit extraction. A comparatively large beam of rectangular cross section is extracted at right angles to the heater filament, from a discharge confined to near the extraction slit (Freeman and Sidenius, 1973).

with a wide variety of ions for long periods of time. The oven shown in Fig. 7.12 is frequently replaced by a small carbon or tantalum crucible which can be inserted through a vacuum lock similar to that shown in Fig. 7.13, Freeman *et al.* (1971). The heat generated from the arc and the massive filament, normally a 2 mm diameter tungsten rod, is frequently sufficient to vaporise material in the small crucible placed close to the filament. The rate of vaporisation can be controlled by moving the crucible towards or away from the filament, Fig. 7.13.

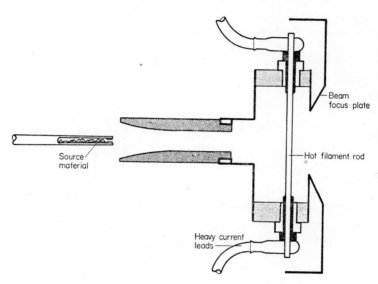

Fig. 7.13. Detail of Fig. 7.12 showing the ion source fitted with crucible containing solid source material. The crucible temperature is changed by adjusting its proximity to the hot filament rod. The crucible can be extracted through a vacuum lock for reloading.

Replacing the crucible by a copper rod capped with the material to be sputtered, Fig. 7.14, and applying a sputtering potential enables the source to be used with materials which have a low vapour pressure or lack thermally stable volatile compounds or are difficult to ionise by other means, such as gold, palladium, iridium, platinum, rhenium and ruthenium. Beams of other refractory metals such as tantalum and molybdenum can be produced by using them as filament rods. An advantage of the rectangular section beam is that it is ideally suited to focussing by a sector magnet without the need of the ion optics normally required with beams of circular section. By careful design good resolution can be obtained.

7.8 MULTIPLY CHARGED ION SOURCES

A considerable effort is at present being put into the development of multiply charged ion sources for high energy heavy ion accelerators, a number of which are described in the Proceedings of the 1972 Vienna Ion Source Conference. Most of the sources are for pulsed operation and are not well suited to ion implantation applications. However, a recent development of the duoplasmatron, Winter and Wolf (1974), as a continuous source capable of producing steady beams of 75 µA of Ar^{3+}, 1·5 µA of Ar^{4+}, 15 µA of Kr^{4+}, 0·5 µA of Kr^{5+}, 1·5 µA of Xe^{6+} and 0·1 µA of Xe^{7+} shows promise as an implantation source. The discharge power required is about 1 kW and 20–50 hours operation times are reported.

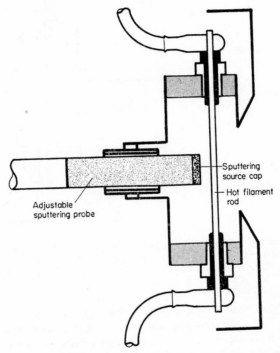

FIG. 7.14. Detail of Fig. 7.12 showing the ion source fitted with a sputtering probe. Voltage is applied to the cap of the probe, attracting ions from the discharge which sputter atoms from the cap, some of which are ionised in the arc discharge and extracted from the source.

There are two modes of operation of the duoplasmatron, Lejeune (1971); the normal mode occurs at a higher pressure and at lower gas pressures a starvation mode which is verging on instability and has hence been little used. The lower pressure also leads to an increased voltage drop in the arc which puts up the probability of multiple ionisation. Winter and Wolf (1974) have made a detailed study of the electronic and magnetic field configurations which produce stable operation for optimum multiply charged ion output. They find current stabilised power supplies necessary, so that gas flow and magnetic field can be varied without affecting the discharge current (normally 5–10 A at about 100 V) and that a much lower magnetic field (< 1 kG) than usual is necessary to stabilise the desired mode. Higher atomic masses require lower magnetic fields, Xe requiring about half the field needed for Ar.

Another interesting point to emerge from their analysis is that a greater beam current for a given gas may be obtained by using a gas mixture than by using the pure gas, a conclusion which can be generalised to other sources. If the additional gas has a higher ionization cross section and/or a higher maximum energy of effective total ionization rate then the number of electrons per gas atom is increased compared with the pure source gas and the discharge will run stably at a lower pressure and the ion output will increase. Even small amounts of Xe added to Ar greatly increase the Ar ion output.

A development of this source to use sputtered metal ions, Winter (1974) has produced triply charged beams of Al, Ti, Fe, Zr, Mo, and W and quadruply charged Ti, Zr, Mo and W.

The disadvantage of multiply charged ion sources, aside from their generally wider energy spread, is their lower output compared with normal ion sources. A disadvantage which present developments are likely to overcome. Their great advantage is their ability to extend the energy range of machines. A 1 MeV accelerator sourced with say Xe^{6+} becomes a 6 MeV machine. This may not be a great advantage for ion implantation but it is for atom location, channelling measurements and Rutherford backscattering. The more favourable charge to mass ratio of the multiply charged ion reduces the deflector plate voltages and the size of the magnets needed to bend the beam, which is a major advantage with vertical machines which require a 90° deflection.

7.9 OTHER ION SOURCES

Many other ion sources have been described using Penning discharges, plasmatrons, sputtering, magnetron discharges, field ionisation, spark ionisation, lasers, exploding wires, mercury cathodes and other methods. Almost any process that can vaporise material and ionise some of it has been tried. We do not propose to discuss them here as numerous works are available to which the reader is referred for further or more detailed information, for example, Wilson and Brewer (1973), Dearnaley *et al.* (1973), Löb and Scharmann (1962), von Ardenne (1956), Rose (1969), Proceedings of the International Ion Source Conference (1969), I.N.S.T.N. Saclay, France and the Proceedings of the Second International Conference on Ion Sources (1972) S.G.A.E. Vienna.

7.10 BEAM HANDLING

Although one normally uses the implant beam to cover a sizeable target area it is essential that it be well focussed so that adjacent elements can be adequately separated and contamination of the implant by the elements next to it in the periodic tables avoided. Resolution of individual isotopes is desirable so that an element can be identified by its natural isotopic abundance. It is not sufficient to tune to the appropriate mass number, as molecular ions and multiply charged ions of different atoms may appear in the same position. Useful information on the operating characteristics of the source also comes from looking at the mass spectrum, obtained by scanning the beam across a slit, or a collector wire across the beam. For example, in using volatile chlorides the relative sizes of the HCl and chlorine peak can indicate the amount of water vapour present which may effect the optimum settings of the source for maximum beam current.

Ions of the same energy but different mass cannot be separated by any electrostatic device. One must use a magnetic field or a combination of electric and magnetic fields or alternating electric fields, as in the quadrupole analyser. The particular choice depends on the geometry required, the characteristics of the beam and the resolution we need. A sector magnet for example is the simplest system but the beam is deflected by an angle equal to the sector angle. A quadrupole mass analyser is a straight through device but a considerable flux of neutrals and metastable atoms may also come through. We will discuss only the sector magnet and the velocity filter below.

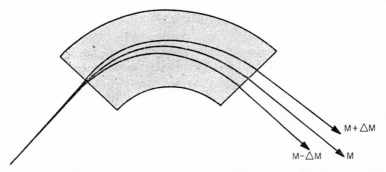

FIG. 7.15. The basic sector magnet mass analyser. For normal incidence on the pole faces of the incident and exit beam of the selected mass the beam deflection is equal to the sector angle of the magnet. Higher masses are less deflected and lower masses more.

7.11 THE SECTOR MAGNET MASS ANALYSER

In its journey from source to target the beam must not only be shaped and focussed, it must be held together. The high current density beams used for ion implantation pose considerable space charge problems, as the positive charges repel each other and produce beam spread. The characteristics of ion optical devices are usually first calculated on the assumption of a tenuous beam, where the fields in the device are not greatly perturbed by the current passing through them. This is not the case in ion implanters. The beam currents are large enough to modify the performance of the optical elements and the beam itself can be quite complex, perhaps even partially space charged neutralised from undergoing collisions with residual gas atoms. So that although the literature on electron optics is vast much of it cannot be applied to heavy current ion beams without considerable reservation.

7.11 THE SECTOR MAGNET MASS ANALYSER

A charged particle moving with constant momentum, p, in a uniform magnetic field, B, follows a circular path. The radius, R, of the path is given by

$$R = \frac{p}{eB} = \frac{1\cdot 44}{B}\left(\frac{MV}{n}\right)^{1/2}$$

where M is the ion mass in atomic mass units, V is the accelerating potential in volts, n the charge state of the ion and B the magnetic field in gauss.

Thus ions of different mass but with the same accelerating voltage and charge state will follow paths of different radius, the radius varying as the square root of their mass, and mass separation occurs.

For the sector magnet shown in Fig. 7.15 the beam deflection for the collimated beam of mass M, incident on and leaving the magnet normal to

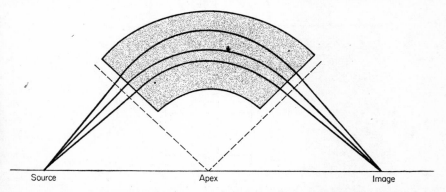

FIG. 7.16. For normal incidence and exit of the beam from a sector magnet the source, magnet apex and image are colinear (Barbers Rule). The dotted line allows for fringing fields which extend from the magnet by a distance of the order of the gap width.

the pole faces, is equal to the sector angle θ and the position of the source and collector are not important. However, all real beams diverge, all sources are of finite size and a sector magnet focusses them in the plane of the diagram. Barber's rule (Enge, 1967) is a useful guide and states that the source, its focussed image and the apex of the magnet lie in a straight line for the beam incident on, and exiting normally from, a sector magnet. The dotted lines in Fig. 7.16, which illustrates Barber's rule, shows an apex of a magnet which is larger than the actual magnet. This is because the field of any real magnet extends outside the magnet gap and the so called Fringing Field can be allowed for by considering the actual pole faces to be outwardly displaced parallel to themselves by an amount which is readily calculated for a known magnet, Enge (1967), Wollnick (1967), (typically equal to the width of the gap).

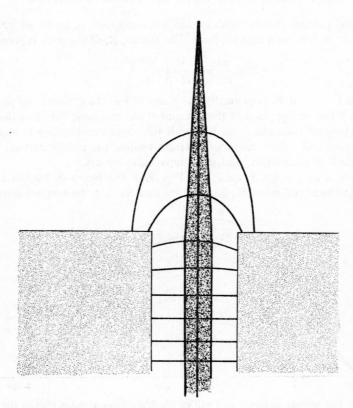

FIG. 7.17. Focussing of the beam produced by the curved magnetic field lines of the fringing field which extend outside the gap of the magnet.

7.11 THE SECTOR MAGNET MASS ANALYSER

A further beneficial effect of the fringing field is shown in Fig. 7.17 which is in a plane normal to that of Fig. 7.16. The bulging magnetic field lines act as a lens and produce a focussing effect in this plane.

The sector magnet, with proper design, can analyse and focus ion beams with a sufficient resolution for ion implantation. A slit source is normally used and no additional ion lenses are required. It is convenient to be able to vary the image position for optimum focus without moving the source and this can be achieved by tilting the magnet pole faces. At non-normal incidence, Herzog (1954), Barber's rule does not apply and the image can be focussed by using adjustable cylindrical entrance and exit edges to the pole faces, Fig. 7.18, which can be moved to produce the desired focus on the target. Such a configuration is used on the Harwell separator which has a 60°, 11 kilogauss sector magnet.

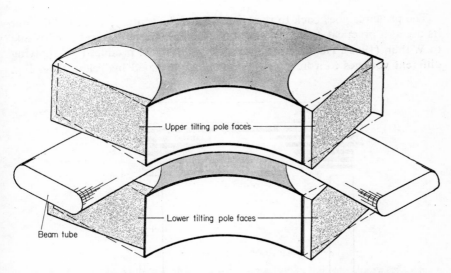

FIG. 7.18. Semicylindrical pole face inserts are shown which can be rotated about an axis through the centre of the pole face normal to the plane of the beam trajectory. This change of angle of incidence changes the focal length of the magnet and the situation shown in Fig. 7.16 no longer applies. The image position can thus be adjusted without moving source or magnet.

The slit source is not essential and many separators use a more complicated system with cylindrical section beams and additional focussing elements. The slit source however does lead to simple scanning of the beam over the target and because of its greater beam current machine time for heavy implant doses is considerably reduced.

7.12 THE VELOCITY FILTER

If a charged particle beam traverses a region where an electric field E and a magnetic field B are perpendicular to the incident beam and to each other it will be undeflected only if its velocity, v, is such that the velocity dependent force, Bev, due to B is equal to and opposite to the velocity independent electrostatic force, eE, due to E, that is

$$Bv = E$$

The velocity depends on the mass M and the accelerating voltage V

$$eV = \tfrac{1}{2}Mv^2$$

whence singly charged ions of mass M will be transmitted for a field ratio of

$$E/B = (2eV/M)^{1/2}$$

The principle goes back to the beginning of the century, Wien (1902), but its use as a practical device in an isotope separator is more recent and is due to Wahlin (1965). He uses a shaped electrostatic field produced by applying different voltages to a number of thin shims insulated by Teflon tape, Fig.

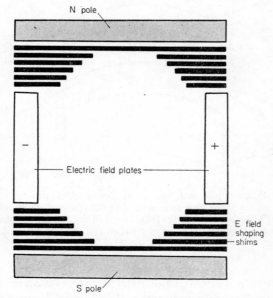

FIG. 7.19. A velocity filter using crossed magnetic and electric fields. The optics of the device are improved by a number of thin metal field shaping plates insulated from each other to which different voltages are applied. The beam passes centrally through the device normal to the plane of the diagram (after Wahlin, 1965).

7.19. This considerably improves the optics of the device and it has since been used in a number of implanters.

The velocity filter has the advantage of an undeflected beam which simplifies the geometry of the machine. This has, however, proved to be a mixed blessing as neutrals from the source, impurity atoms in the vacuum system and the desired atoms which may have undergone charge exchange with loss of energy at various points in the system are all swept onto the target and may effect our implantation results. They can however be removed by deflecting the beam just before it reaches the target.

7.13 BEAM SCANNING

The most common implantation requirement is for uniform coverage of a few square centimeters. One cannot rely on adequate beam cross sectional uniformity using a stationary beam defocussed to cover the area. Some form of scanning and focussing is in any case necessary for beam identification and control and it is a simple extension to extend the scanning to cover the region required.

With a sector magnet separator, Fig. 7.16, the image of a slit source can be scanned by simply sinusoidally modulating the accelerating voltage and selecting the centre of the sweep. Better uniformity and beam utilisation can be had with a linear sweep voltage. Collecting the beam behind a narrow slit placed at the focus can be used to display on an oscilloscope, the isotopes of the implant ion collected Fig. 7.20. From an isotopic distribution chart, identification and the location of impurities and doubly charged species of different mass is straightforward, the separator acting as a mass spectrometer. When the beam and scanning system are satisfactorily adjusted the slit is replaced by the specimen target to be implanted.

If a large area is to be implanted the scan may be so great that overlap by adjacent isotopes may occur over some of the target. This is more likely at high mass numbers as the dispersion of most analysers is inversely proportional to the mass of the ion used. A second scanning system placed behind a slit at the focus reduces this trouble. The slit at the focus selects only the required mass and the second scanning system spreads it over the target.

For beams of circular cross section $X-Y$ scanning must be used, preferably with sawtooth voltages applied to orthogonal deflecting plates. Care must be taken to avoid harmonic frequency relations between the X and Y frequencies and between them and the mains which may give rise to a non-uniform dose pattern over the surface. Since, at best, the beam profile is Gaussian sufficient overlap between individual traces and at the edge of the specimen must be allowed.

Moving wire scanners are commonly used, one giving the beam profile in a vertical and another giving the profile in a horizontal direction. This aids beam focussing and helps to ensure that adequate, but not wasteful overlap is used.

Simple scanning systems can vary the angle of incidence by several degrees. The resultant small change in penetration depth is acceptable for many purposes but the angular spread is orders of magnitude outside the requirements of a channelling implant. The precision needed for channelling can be achieved by a "dog leg" system using two coupled scanners in series to sweep the beam while keeping the angle of incidence on the target. If the long term stability of the separator is adequate the targets can be scanned through a fixed beam but in general this is a more complex and expensive solution if precision is required.

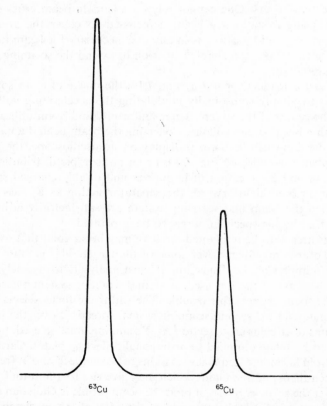

FIG. 7.20. Mass spectrum of a beam of natural copper ions showing the two isotopes at 63 and 65 a.m.u.

7.14 TARGET MANIPULATION

For most research applications, such as channelling and Rutherford backscattering, accurate target alignment is necessary and remote control three axis goniometers are the most convenient. A widely used design due to Knox (1970) Fig. 7.21 uses straightforward worm drive and vacuum stepping motors to achieve a setting accuracy of about a fiftieth of a degree. The large

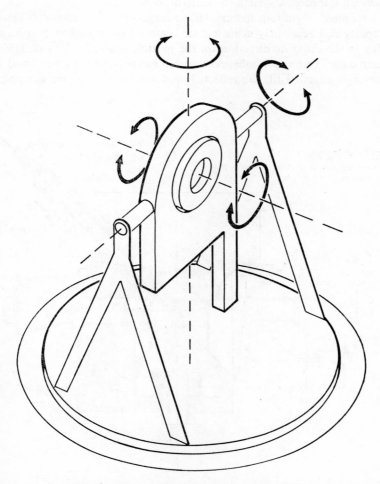

FIG. 7.21. A goniometer using worm drives and precision gears with stepping motors mounted in the vacuum chamber (after Knox, 1970). The rotation axes are indicated but the stepping motors have been omitted from this figure.

target ring enables a heating stage to be installed and observations can be made in reflection or transmission.

A more precise two axis skewed goniometer capable of a precision of better than 0·005° on both axes is shown in Fig. 7.22. The goniometer is steplessly hydraulically driven and the angles measured to the above precision with Moire optical precision angle detectors, Dalglish (1974), which communicate directly in 16 bit binary to a PDP 11 computer. The computer can rapidly set the required angle by simultaneous manipulation of the two axes through a feedback system to the hydraulic controls.

The alignment of smooth flat crystalline targets in the goniometer can be done rapidly and accurately using a laser beam set to be collinear with, and travelling in the opposite direction to, the particle beam, Price *et al.* (1973). The beam enters through a hole in a screen a metre from the crystal and the goniometer is adjusted till the reflected spot seen on the screen disappears

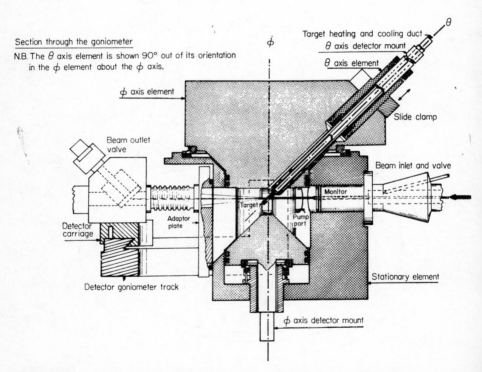

FIG. 7.22. Skew axis goniometer with a precision of 0·005° about both axes. External hydraulic drive and optical digital angle measurement for direct computer interfacing (after Dalglish, 1974).

down the entry hole. The crystal surface is then normal to the laser beam. A rotation of 180° about any axis perpendicular to the beam brings the crystal surface normal to the particle beam. If the centering of the reflected spot is good to only 2mm the orientation of the crystal surface is to within 0·05° and this can easily be improved if necessary. The method is particularly suited to targets with well defined cleavage faces, such as the alkali halides which cleave on (100) surfaces but is useful for any target whose crystallographic orientation relative to a flat smooth surface is known. The extension of this use of the optical lever to accurately and simply measure small angular displacements of crystals being implanted or channelled does not seem to have been made.

For most routine implants high angular precision is not needed and a larger target holder which enables many specimens to be mounted and processed without breaking the vacuum may be more desirable. A number of such chambers have been described, Dearnaley et al. (1973).

7.15 BEAM TRANSPORT

The control and focussing of charged particle beams is a highly developed branch of engineering as evidenced for example by the precision achieved by modern electron microscopes. The analytic methods are usually based on paraxial ray optical analogues and the aberrations described analagously to optical aberrations. These methods predate the computer and were developed for ease of computation. In the absence of space charge and with the paraxial approximation, linear ion optical elements can each be represented by a transformation matrix which relates the radius and shape of the trajectories into and out of the element. The matrices for specific elements are given in Wilson and Brewer (1973), together with a discussion of the method. The matrix multiplication involved in tracing the beam through the beam optics is normally done by computer which allows easy adjustment of the element parameters to produce the observed image on the target.

With an ion source producing beams of the order of a milliamp the paraxial and zero space charge assumptions of the matrix method break down, particularly near the extraction system when the beam density is high and the ions are moving slowly. Recourse must be had to computer trajectory tracing methods with the beam region broken up into cells each of which has specified fields. The smaller the cells the slower the calculation and the greater the demands on the computer memory but the more accurate is the result. Some economy can be achieved by taking large cells in regions where the field gradients are low and smaller cells for steep gradient regions. The procedures are well established and widely used, Steffan (1965), and there is much to be said for the view that we should stop pretending that ions and electrons are photons and give up forcing particle beam calculations into optical formalisms which require numerous aberration coefficients to correct the results, and which are much less successful for charged particle beams than they are for light optics.

The computer can be programmed to output the data graphically as stereoscopic perspective pairs which when viewed through a stereoviewer give an instantly understandable feel of the effects of beam line modifications, Dalglish and Kelly (1974).

Even with the detailed calculations there remain effects due to edge scattering at apertures, collisions with residual gas and wall effects. It is axiomatic that beams should be prevented from line of sight interaction with insulating surfaces as the insulators charge up in a non-reproducible way and the resultant electrostatic fields degrade the beam quality. If the beam, or secondary particles from it, strikes insulators residual hydrocarbon vapour is cracked on their surface and eventually produces conducting carbon layers, leading to electrical breakdown and limiting their life. The usual procedure is to use metallic shields or a reentrant shape for accelerating electrodes, Septier (1967). In a separator the problems are greater than in a normal accelerator using light gas ions. The general purpose separator is used for a variety of substances in turn and the inner wall surfaces, which are initially conducting, can become inhomogeneously insulating, particularly if the ions of metals which oxidise easily are used. Chlorine, a common element in many ion source feed materials, can also attack the metal walls and reduce their conductivity. The proper use of removable shields and regular cleaning are required to reduce the contamination problem to a reasonable level.

7.16 THE MEASUREMENT OF DOSE

In principle dose measurement is simple. The beam currents range from a nanoamp to a milliamp which is well within the capacity of normal current integrators, and all that is required is to connect the target to the integrator which is set to interrupt the beam at the required total charge delivered to the target, usually with a shutter. As anyone who has used a Faraday cup knows this procedure can be relied upon to give the wrong answer, unless one uses a suitable arrangement of biassed plates to suppress secondary and tertiary electrons, because these will be registered by the integrator as positive ions. Ions will also leave the surface but they are much fewer in number than the electrons and will give a smaller error. The number of secondary electrons per incident ion is a function of the ion species and energy and the nature of the surface. The problems are compounded for insulators whose emission of secondary particles is much higher and which are subject to surface charging. It may be possible to implant through a thin conducting layer, at the expense of risking contamination but the more usual solution is to flood the surface with electrons from an adjacent emitter which needs to be well stabilised and included in the current integrator circuits as the electron flood current may be many times the small ion beam current. As the implantation proceeds the surface is changed from the pure target material to a mixture of target and implant atoms with a corresponding change in work function and hence in secondary electron emitting characteristics. As we cannot always reliably correct for the loss of secondary electrons they must be suppressed. Electrostatic suppression is frequently used with negatively biassed plates in front of the target; the larger the target area the less effective this procedure becomes. Sputtered positive ions will be attracted to the negatively charged suppressor and may in turn create secondaries. If sputtering is considerable, as when both target and implant atom are massive, the suppressor current may have to be monitored and the dose current integrator corected accordingly. Another solution is to make the entire scattering chamber a Faraday cup and insulate it from the system. Stray capacitance effects at small beam current can be troublesome with this system.

Magnetic suppression, with a field of about a hundred gauss normal to the ion beam at the target surface, is also effective. The electrons spiral along the field lines and can be easily collected. There remains the problem of neutrals which contribute to the dose but are not counted by the integrator and electrons swept along with the beam which do not contribute to the dose but are counted. The neutrals may be minimised by a beam deflection near the target and by maintaining the best possible vacuum in the machine. Little can be done about the electrons in the beam unless one knows the beam content and profile. A number of profiling devices are available which cyclically move wires through the beam and measure the current collected on the wire or scattered from it. They are used on many machines because of their simplicity and reliability. Their response, however, depends on the beam species and energy and, like the target, will change with surface contamination. They are not absolute and must be calibrated for each operating condition. Nor can they distinguish between electrons and ions in the same beam.

A device which can measure beams down to 10^{-12} A and has the capacity to both profile the beam and distinguish between ions and electrons has recently been described, Fig. 7.23 Dalglish and Kelly (1972) but has not yet been applied to separator beams. The detecting head of this ion beam monitor responds only to the field produced by the passing charge and hence may be absolutely calibrated by electrons passing through it on a wire. Its calibration is of course independent of the energy and species of ion beam measured.

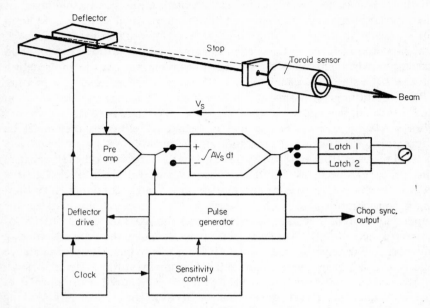

Fig. 7.23. A block diagram of the Dalglish and Kelly (1974) ion beam monitor. A toroidal beam detector integrates the voltage pulse produced by a pulsed undeflected beam or momentarily deflects a steady beam change the flux linked with the toroid. Direct absolute calibration is possible by using a mimic pulse passed through the toroid on a coil. The sensitivity is varied by varying the number of integrators added with a minimum limit of 10^{-12} A. The beam profile can also be extracted.

Even if we can make full allowance for electrons emitted, there still remains the problem of sputtering which always occurs when heavy ions strike a surface. Not only are implanted ions resputtered along with target ions, thus reducing the total dose left in the target, but this implantation profile may be seriously modified. Most sputtered ions are neutral so the departure of the implant ions will not be registered by our current integrator and we will record a larger dose than we have retained in the target. Sputtering yield depends on the projectile target mass ratio and will in general be different for matrix and implant. If the sputtering yields are high enough a saturation distribution should be reached, Carter, Colligon and Leck (1962), Krimmel and Pfleiderer (1973) but the uncertainties in the sputtering yields of many materials and their variations with energy and angle of incidence frequently present an exact correction for these effects. At higher energies where the implant ion and most of its radiation damage is deposited well below the surface the sputtering effects are much reduced.

REFERENCES

Almén, O. and Nielsen, K. O. (1957). *Nucl. Instr. and Methods* **1**, 302.
Brewer, G. R., Currie, M. R. and Knechtili, R. C. (1961). *Proc. I.R.E.*, **49**, 1790.
Carter, G., Colligon, J. S. and Leck, J. H. (1962). *Proc. Phys. Soc.* **79**, 299.
Dalglish, R. L. and Kelly, J. C. (1972). "Proc. Second Int. Conf. on Ion Sources."
Dalglish, R. L. (1974). To be published.
Dalglish, R. L. and Kelly, J. C. (1974). To be published.
Dearnaley, G., Freeman, J. H., Nelson, R. S. and Stephen, J. (1973). "Ion Implantation", North-Holland.
Enge, H. A. (1967). *In* "Focusing of Charged Particles" (Ed. A. Septier), Academic Press, New York.
Freeman, J. H. (1963). *Nucl. Instr. and Methods*, **22**, 306.
Freeman, J. H., Gard, G. A. and Temple, W. (1971). A.E.R.E. Research Report R6758.
Freeman, J. H. and Sidenius, G. (1973). *Nucl. Instr. and Methods*, **107**, 477.
Hague, C. A. and Donaldson, E. E. (1963). *Rev. Sci. Instr.* **34**, 409.
Herzog, R. (1954). *Zf. Physik*, **89**, 447.
Illgen, J., Kirchner, R. and Schulte in den Bäumen, J. (1972). *IEEE Trans. Nucl. Sci.* NS-19 no 2, 35.
Johnson, P. G., Bolson, A. and Henderson, C. M. (1973). *Nucl. Instr. and Methods*, **106**, 83.
Knox, K. C. (1970). *Nucl. Instr. and Methods*, **81**, 202.
Komrov, V. L., Tsepakin, S. G. and Chemyakin, G. V. (1969). "Proc. Int. Conf. on Ion Sources", I.N.S.T.N., Saclay.
Krimmel, E. F. and Pfleiderer, H. (1973). *Rad. Effects*, **19**, 83.
Langmuir, I. and Taylor, J. B. (1933). *Phys. Rev.* **34**, 423.
Lejeune, C. (1971). Thesis, Paris.
Lerner, J. (1965). *Nucl. Instr. and Methods*, **38**, 116.
Löb, V. H. and Scharmann, A. (1962). *Kerntechnik*, **4**, 1.
Magnuson, G. D., Carlston, C. E., Mahaderan, P. and Comeaux, A. R. (1965). *Rev. Sci. Instr.* **36**, 136.
Masic, R., Sautter, J. M. and Warnecke, R. J. (1969). *Nucl. Instr. and Methods*, **71**, 339.
Nielsen, K. O. (1957). *Nucl. Instr. and Methods*, **1**, 289.
Nielsen, K. O. (1969). "Proc. Int. Conf. on Mass Spec." Kyoto, Japan.
Price, P. B., Hollis, M. J. and Newton, C. S. (1973). *Nucl. Instr. and Methods*, **108**, 605.
Rose, P. H. (1969). "Ion sources for heavy ion accelerators", High Voltage Engineering, Burlington, Mass.
Septier, A. (Eds) (1967). "Focusing of Charged Particles", Academic Press, New York.
Shimizu, K., Kawakatsu, K. and Kanaya, K. (1973). *Nucl. Instr. and Methods*, **111**, 525.
Sidenius, G. and Skilbreid, O. (1961). "Electromagnetic Sepn. of Radioactive Isotopes", (Eds Higatsberger and Viehböck), Springer-Verlag, Vienna.
Sidenius, G. (1969). "Proc. Int. Conf. on Ion Sources", Saclay, p. 401.
Smith, M. and Langmuir, I. (1926). *Phys. Rev.* **28**, 727.
Steffan, K. G. (1965). "High Energy Beam Optics", John Wiley, New York.

Thonemann, P. C. (1946). *Nature,* **158,** 61.
Thonemann, P. C., Moffatt, J., Roaf, D. and Saunders, J. H. (1948). *Proc. Phys. Soc.* **61,** 483.
von Ardenne, M. (1956). "Tabellen der Elektronenphysik und Ubermikroskopie", Veb. Deutscher Verlag der Wissenschaften, Berlin.
von Ardenne, M. (1961). *Expt. Tech. Physik,* **5,** 227.
Wahlin, L. (1965). *Nucl. Instr. and Methods,* **38,** 133.
Wien, W. (1902). *Ann. Physik.* **8,** 260.
Wilson, R. G. and Brewer, G. R. (1973). "Ion Beams", John Wiley, New York.
Wilson, R. G. and Jamba, D. M. (1967). *J. Appl. Phys.,* **38,** 1976.
Winter, H. (1974). "Proc. 7th Yugoslav Symposium on Physics of Ionized Gases", p. 79.
Winter, H. and Wolf, B. H. (1974). *Plasma Physics,* **16,** 791.
Wollnick, H. (1967). *Nucl. Instr. and Methods,* **53,** 197.

CHAPTER 8

Analysis Techniques with Ion Beams

8.1 INTRODUCTION

In this chapter we will discuss the applications of ion beams as a means of characterising surfaces rather than for inducing property changes. For analysis, the energy used may span right across the range used for ion implantation depending on the type and quality of information that is required. A few tens of eV are sufficient for some of the electronic processes yielding valence band (chemical) information for the outermost monolayers, whereas, kilo-electronvolt photons are required for X-ray photoelectron spectroscopy, or heavy ions at keV energies to produce ion stimulated X-rays. For Rutherford backscattering and nuclear reaction techniques MeV beams of light ions are used, from which the position of atomic nuclei over depths of a few microns can be obtained. The only unifying feature of these analytical techniques is that charged particles are used as probes. There has been a very great expansion recently in the applications of modern analytical techniques and we will confine ourselves in this chapter to a brief survey consistent with studies on ion implanted and sputtered surfaces. Fortunately, there are now excellent review articles dealing with specific topics, for example Nicolet *et al.* (1972), Wolicki (1972), Bird *et al.* (1974), Rivière (1975); there are also conference proceedings (Mayer and Ziegler, 1974) and books (Kane and Larrabee, 1974), to which the reader is referred for further information. In this part of this book Rutherford backscattering and related MeV processes will be discussed in rather greater detail than other techniques because RBS is still the most frequently employed method of examining ion implanted surfaces.

8.2 RUTHERFORD BACKSCATTERING

In the classic experiments of Rutherford and his co-workers in 1911 the elastic scattering of light energetic ions was first reported. As an analytical tool though, Rutherford backscattering did not emerge until 1957 when Rubin, Passell and Bailey discussed the application of elastic scattering to elemental surface analysis. Ten years later the famous experiment of Turkevich, Franzgrote and Patterson (1967) was reported in which an alpha source was directed at the moon's surface from a Surveyor spacecraft to perform the first direct elemental analysis of the moon. In recent years the application of Rutherford backscattering has increased enormously, particularly in the semiconductor field where technological demands have accelerated the need for a technique capable of detecting foreign atoms located in the first few microns of the surface of an otherwise pure material. Rutherford backscattering experiments frequently form an integral part of papers on ion implantation as can be judged by consulting conference proceedings which have appeared in the last few years (e.g. Palmer *et al.* (1970), Crowder (1973), Picraux *et al.* (1974), Datz *et al.* (1975)).

In principle the experimental arrangement is straightforward and merely involves the collection and analysis of high energy particles rebounding from a target specimen after part of an impinging beam has undergone billiard ball collisions with its constituent atoms. The electronics associated with backscattering has to sort a rapid stream of pulses into energy groups, and several proprietary units are now available to do this. Typical backscattering equipment is shown in Fig. 8.1 and in Appendix III a block diagram is given of the electronic components. The specimen to be analysed is exposed to a monoenergetic beam of light ions (such as protons or helium ions) at currents $\sim$ 10–30 nA and with an energy typically of 2 MeV. A high degree of collimation is not required, although it is important to have a well defined beam for channelling experiments (see Section 8.3). Energies above 2 MeV are usually used if resonant reactions are to be exploited or a greater depth of sampling by the probing beam is required. The beam size is usually about 1 mm^2 although it is possible to produce beams of micron dimensions using a specially designed fine focus quadrupole lens arrangement (Cookson and

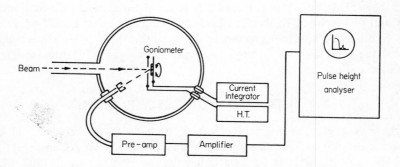

FIG. 8.1. Schematic diagram of backscattering apparatus.

Pilling, 1970). The target chamber is evacuated to at least 10^{-5} Torr and the particles which recoil from the specimen at angles approaching 180° with respect to the incoming beam penetrate a silicon surface barrier detector where they lose energy by creating ionization. The detector is a large area diode consisting of an extremely thin *p*-type layer on one side of a high

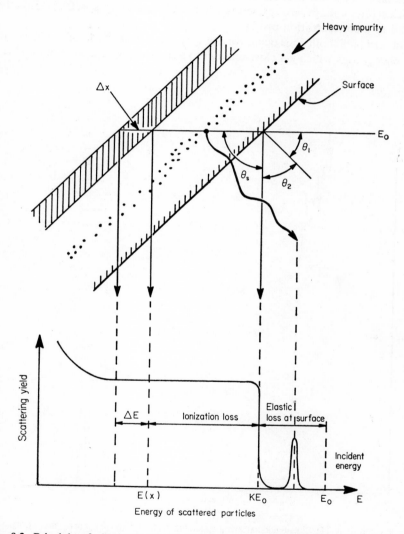

FIG. 8.2. Principle of elastic backscattering, shown for an idealized sample of a light material containing a heavy impurity. θ_s is the scattering angle (after Bøgh, 1968).

purity *n*-type wafer. For every incident ion, the detector acts as a very rapid ionization chamber, producing a voltage pulse which is proportional to the energy of the backscattered ion. The pulses are amplified and stored in a pulse-height analyser. Since the energy of the backscattered particles is related to the mass of the struck atom and the number of collisions is proportional to the number of scattering centres, we may, by sorting the scattered particles into voltage groups, establish a backscattering spectrum directly related to a concentration profile.

"Billiard ball" collisions occur for the very small proportion of the incident high energy ions which get close enough to a nucleus, within 10^{-4} or 10^{-5} Å,

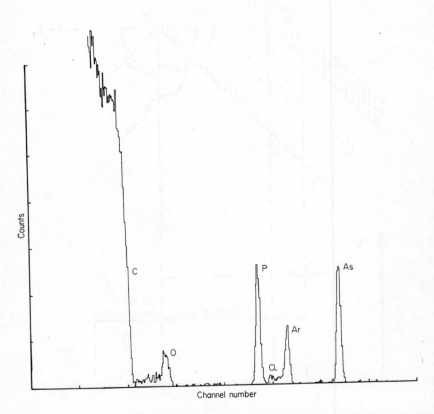

FIG. 8.3. 2·0 MeV He$^+$ ion backscattering spectrum from a vitreous carbon disc containing a thin layer of implanted and adsorbed elements. Illustrating the separation of different species due to the kinematic factor K.

to undergo violent elastic scattering due to the Coulomb electrostatic repulsion, which is an elastic scattering event when both energy and momentum are conserved. Taking firstly the case of collisions at the surface of a target, the energy of rebound will be a simple function of the masses of the incident and target ion. The probability of interaction (the cross section) is related to the atomic numbers of the target and incident ion, and the energy of the incident beam. The energy of the particles scattered from the surface is

$$E = KE_0 \tag{8.1}$$

where

$$K = \left(\frac{M_1 - M_2}{M_1 + M_2}\right)^2 \tag{8.2}$$

for backscattering through 180°, and the quantities are defined as in Fig. 8.2. At a scattering angle θ_s (Fig. 8.2) the kinematic factor is reduced such that

$$K = \frac{M_1^2}{(M_1 + M_2)^2} \left[\cos \theta_s + \left(\frac{M_2^2}{M_1^2} - \sin^2 \theta_s\right)^{1/2}\right]^2. \tag{8.3}$$

We therefore have that for a fixed detector position the energy of the scattered ion depends on the mass of the struck atom. Thus for very thin targets, individual elements may be identified (Fig. 8.3).

As the beam penetrates further into the target it suffers energy loss by inelastic ionization losses (Chapter 2), but to consider first the simpler case of targets which are thin compared with the penetration range of the incident beam, the number of particles S scattered by atoms of mass M_2 into solid angle ω is given by

$$S = n N(M_2) \Delta x \, \omega \, \sigma (E_0, \theta_s, M_1, Z_1, M_2, Z_2) \tag{8.4}$$

where n = number of incident beam atoms

$N(M_2)$ = number of scattering centres of mass M_2 per cm^3

Δx = target thickness in cms.

Thus for a very thin target consisting of multielement layers, such as discussed in Chapter 5, the area under each of the peaks is proportional to the number of atoms of each type in the specimen and its scattering cross section.

The Rutherford cross section is simply related to Z_1, Z_2, the incident beam energy E_0 and the scattering angle θ_s by

$$\sigma \propto \left(\frac{Z_1 Z_2}{E_0}\right)^2 \frac{1}{\sin^4(\theta_s/2)}. \tag{8.5}$$

When E_0 is expressed in MeV the differential cross section is given by

$$\left(\frac{d\sigma}{d\omega}\right)_{cm} = 1 \cdot 296 \left(\frac{Z_1 Z_2}{E_0}\right)^2 \left(\frac{M_1 + M_2}{M_2}\right)^2 \frac{1}{\sin^4(\theta_s/2)} \text{ mb/sr} \tag{8.6}$$

for particles of mass M_1, charge $Z_1 e$ striking target atoms of mass M_2, charge $Z_2 e$. The value of $(d\sigma/d\omega)$ for a particular combination of Z_1, Z_2 and E_0 may therefore be calculated with a high degree of accuracy for a given scattering angle. There is however, a small correction (about 4% at laboratory scattering angles of 164°, assuming the primary ion is helium) which should be applied to the differential cross section to allow for recoil of the target atoms. Clearly this is more important for light targets (e.g. below Si). From kinematical considerations the centre of mass scattering angle θ is related to the laboratory scattering angle ψ by

$$\theta = \psi + \sin^{-1}\left(\frac{M_1}{M_2}\sin\psi\right) \tag{8.7}$$

and accordingly

$$\left(\frac{d\sigma}{d\omega}\right)_{lab} = 1 \cdot 296 \left(\frac{Z_1 Z_2}{E_0}\right)^2 \left[\operatorname{cosec}^4\left(\frac{\psi}{2}\right) - 2\left(\frac{M_1}{M_2}\right)^2 + \ldots\right] \text{ mb/sr} \tag{8.8}$$

where the notation is as before. Tables of Rutherford cross-sections are available for fixed energy and detection angle (for example, Wolicki, 1972). Values for $\operatorname{cosec}^4(\psi/2)$ for given ψ and the full laboratory and centre of mass kinematical and geometrical relationships can be found in the reference book by Marion and Young (1968).

The preceding equations show that from the relatively simple relationships (8.1), (8.2) and (8.5) a basis for mass analysis exists. However, the resolution between adjacent masses decreases as the mass of the target atom increases because of the resolution of the detector and the dependence of K (equation (8.2)) on the mass difference squared. This means that for progressively higher masses, the energy of ejection of a backscattered particle becomes less distinguishable from that arising from a species very close in mass. On the other hand, the rapid rise in sensitivity for the high mass elements—due to the rise in Rutherford cross section (equation (8.5)) means that extremely low concentrations of heavy elements can be detected. Figure 8.4 shows the presence of a few ppm of Pb on a piece of carbon exposed to the atmosphere. In this case the spectrum shows no continuum from the carbon disc because backscattering was carried out with nitrogen ions at 3·5 MeV to eliminate the background. Clearly, backscattering will not occur from those species for which $M_2 \leqslant M_1$. A second advantage of using a higher mass probing beam is that the resolution between adjacent masses in the target is improved which is the principal reason why helium ions are used in backscattering in preference to protons.

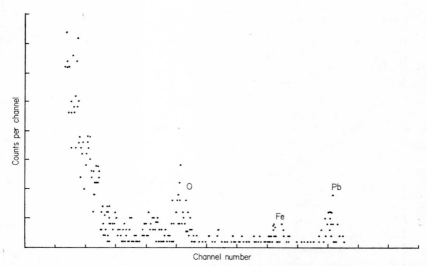

FIG. 8.4. 3·5 MeV N^+ backscattering spectrum of a fraction of a monolayer of Pb deposited from the atmosphere onto a lightly waxed carbon substrate. The other prominent peaks are due to Fe and O contamination, also from air-borne adsorbates (Dearnaley and Squires, 1975).

It is not only mass analysis which makes Rutherford backscattering a particularly versatile tool, since the energy loss of the beam when related to the stopping power of the target can be used for depth analysis over the penetration range of the beam. Since MeV light ions penetrate to a depth of microns, the depth sensitivity of backscattering analysis places it in an analytical range of particular interest to the metallurgist, solid state physicist and the corrosion specialist to give what has been termed "mass-sensitive depth microscopy" (Nicolet et al., 1972).

In essence, the inelastic energy loss (Chapter 2) suffered by the beam means that at progressively deeper levels the "incident" energy seen by a scattering centre is diminished by an amount proportional to the stopping power of the medium. Thus by knowing the stopping power, or electronic energy loss, of the target from consultation with tables (Northcliffe and Schilling, 1970) or a recent compilation of experimental data (Ziegler and Chu, 1973) the depth at which a particular elastic scattering event took place may be calculated. This is the basis for the determination of implantation profiles by Rutherford backscattering and is most readily suited to heavy dopants within a light substrate since under these conditions the backscattering peak is separated from the continuum, Fig. 8.5.

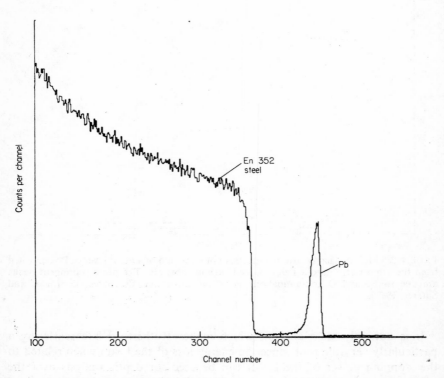

FIG. 8.5. 2·0 MeV He$^+$ backscattering spectrum from En 352 casehardening steel implanted with 6·3 × 10^{16} ions/cm^2 Pb$^+$ at 175 keV.

An early treatment of the thick target case is given by Bøgh (1968) and can also be found in the book by Mayer, Eriksson and Davies (1970). Referring to Fig. 8.6, the particles scattered at a depth x (measured along the beam) have also lost energy by ionization and excitation of the target atoms before and after elastic scattering, at a rate determined by the stopping

8.2 RUTHERFORD BACKSCATTERING

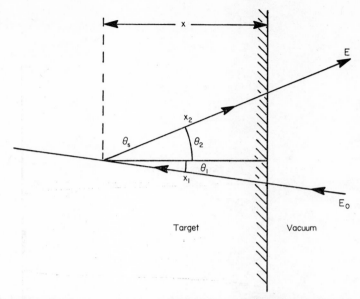

FIG. 8.6. Schematic diagram of parameters used in converting an energy scale into a thickness scale with reference to equations 8.9 and 8.10 (after Mayer et al., 1970).

power $S(E)$ (Bøgh, 1968). Their energy is therefore

$$E(x) = K \left[E_0 - \int_x^0 S(E) \, dx \right] - \int_0^{x \cos \theta_1 / \cos \theta_2} S(E) \, dx. \qquad (8.9)$$

For depths of 1 or 2 microns over which the stopping power is approximately linear with depth we have

$$E(x) \doteq K E_0 - x \left[S(E_0) K + S(E) \frac{\cos \theta_1}{\cos \theta_2} \right] \qquad (8.10)$$

$$x = \frac{K E_0 - E(x)}{\left[S(E_0) K + S(E) \dfrac{\cos \theta_1}{\cos \theta_2} \right]} \qquad (8.11)$$

with E, E_0 in MeV, $S(E)$ in MeV/(mg cm^{-2})

$$x = \left(\frac{K E_0 - E(x)}{\left[S(E_0) K + S(E) \dfrac{\cos \theta_1}{\cos \theta_2} \right]} \right) \frac{10}{\rho} \qquad (8.12)$$

where ρ is the density in gm/cm^3.

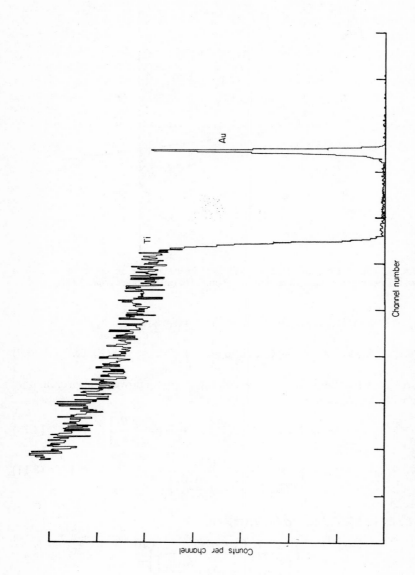

FIG. 8.7. Typical calibration spectrum obtained from a thin evaporated layer of gold on titanium, used in determining instrumental resolution and energy calibration on the horizontal axis.

The horizontal scale is therefore simultaneously a mass scale and a depth scale, depth increasing to the left and mass to the right. Consequently it is necessary for element identification and depth analysis to know the energy of the backscattered particles which fall within a particular channel on the spectrum, i.e. the horizontal scale has to be calibrated. This is most easily done by backscattering for a few minutes from a target consisting of a thin evaporated layer of Au on a Ti substrate. The characteristic spectrum from this specimen consists of a sharp peak corresponding to the gold, well separated from the lighter substrate (Fig. 8.7). Knowing the kinematical scattering ratios for Au and Ti (equation (8.3)) it is a simple matter to arrive at a calibration in terms of keV/channel for the given electronic settings (which are then unaltered). The resolution of the detector, which principally

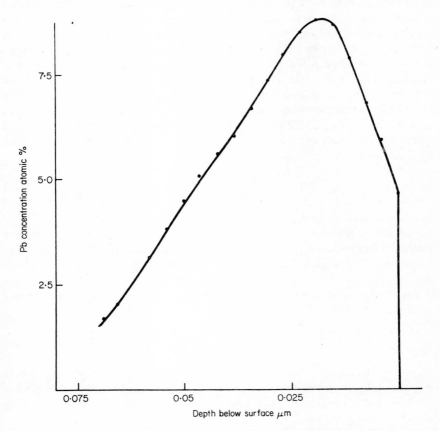

FIG. 8.8. Depth distribution of 175 keV Pb$^+$ ions in En 352 steel calculated from backscattering spectrum given in Fig. 8.5.

determines depth resolution in backscattering, may also be obtained from such a calibration spectrum. If the FWHM of the Au peak, for the example cited above, is 4 channels then at 5 keV/channel the resolution is 20 keV, corresponding to about 300 Å in metallic targets. The resolution may be improved by cooling the detector.

The procedure for obtaining the concentration profile, Fig. 8.8, of Pb implanted into steel from a backscattering spectrum (Fig. 8.5) is most readily achieved by subdividing the "dopant" peak into convenient groups of channels on the horizontal scale, each of which has a mean scattering energy E_i and stopping power $S(E_i)$. By applying equation (8.11) the real depth of each elemental strip can be calculated, while the number of scattering centres (i.e. implanted ions) located at this depth is directly related to the number of counts divided by Z_2^2. Values of the stopping power over incremental depths in many targets are not available however, and it is frequently necessary to extrapolate from quoted values for materials of similar composition. For the case of implanted metals, which the absolute concentration of ions is rarely more than a few percent, it is sufficient for relative assessments of concentration to take the electronic energy loss of the pure material. For interdiffused films, or layers of corrosion products, for example, for which a substantial proportion of the target material may constitute a second element, it is necessary to calculate a proportional stopping power based on the values for each pure material. In this case, $S(E)$ in equation (8.11) is replaced by $\bar{S}(E)$ where

$$\bar{S}(E) = \frac{x}{x+y} S_A(E) + \frac{y}{x+y} S_B(E) \tag{8.13}$$

for the compound $A_x B_y$.

The energy loss per unit length (dE/dx) is frequently the most convenient method of expressing the stopping power of a medium. However a more fundamental concept is that of energy loss based on the number of atoms per square centimetre that are traversed by the beam (Chu et al., 1973). Thus if ρ is the density of the target and M its atomic mass then the stopping cross section ε is

$$\varepsilon = \frac{1}{N} \frac{dE}{dx} = \frac{M}{N_0} \frac{dE}{dx} \frac{1}{\rho}, \tag{8.14}$$

where N_0 is Avogadro's number. Stopping cross sections expressed in this way (as opposed to the stopping power $S(E)$) are then linearly combined for a compound material since the cross section then refers to an areal density of specific composition (Feng et al., 1973; Meyer et al., 1973). Thus

$$\varepsilon_{Fe_3O_4} = 3\,\varepsilon_{Fe} + 4\,\varepsilon_O. \tag{8.15}$$

Although calculations associated with backscattering appear complicated, the equations are simple and it is required only to adopt a systematic arithmetical procedure which in practice is rapidly attained. It is frequently possible to deduce the major features of a given sample from inspection of the backscattering spectrum, as is discussed in the cases below. Worked examples of the concentration of species in interdiffused metallic films are given in review papers by Wolicki (1972) and Chu et al. (1973). To arrive at absolute values of the amount of implanted species it is necessary to measure the detector solid angle (equation (8.4)) and the accuracy of the result will depend on the precision to which this can be done as well as the electronic resolution. A convenient scheme which makes use of a backscattering standard has been initiated by Baglin (1975). A set of nominally identical silicon specimens into which a precisely known amount of bismuth has been implanted to a shallow depth are now available. The backscattering data obtained from such a specimen can be used to establish the resolution of a given chain of electronics, or from a simple comparison with other samples for which the $(1/Z^2)$ dependence of the cross section is included, the absolute amount of any conveniently measured species can be determined in subsequent backscattering experiments for a given detection geometry.

Backscattering has to date found most application in the study of sputter-deposited and evaporated thin films on such substrates as silicon. In addition

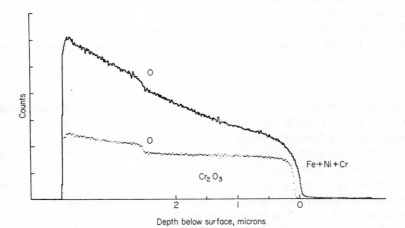

FIG. 8.9. 2·9 MeV He$^+$ backscattering spectrum from a specimen of Type 316 steel corroded in a solution containing NaOH. The spectrum is compared with a powdered standard sample of Cr_2O_3 in order to arrive at the stoichiometric (metal/oxygen) ratio of the oxide film. The oxide exists up to the surface and the depth scale (plotted using the metal edge as origin) illustrates the depth of sampling of the beam (Dearnaley et al., 1975). At 2·9 MeV the elastic cross-section for α scattering from ^{16}O is ∼ 35% greater than Rutherford (Figure 8.17).

to the references already cited, examples of interdiffusion studies in such substrates can be found quoted by Mayer and Tu (1974) and contributions to conferences (e.g. Campisano *et al.*, 1974; DeBonte *et al.*, 1974). Ziegler and Baglin (1971) describe the application of Rutherford backscattering to the determination of concentration profiles of impurities in solids. The determination of stoichiometric ratios in oxide films on metals has recently been achieved by the use of backscattering (Dearnaley *et al.*, 1975); and an example of the copper–oxygen system had been discussed already in Section 5.5. The fundamental difficulty in the examination of oxide/metal systems is the relatively low yield of oxygen with respect to the metal scattering centres because of the Z^2 dependence of the Rutherford cross section. This means that the concentration of oxygen counts is usually masked by the continuum, unless an energy region is chosen where the cross section is higher than Rutherford, or thin specimens are used, Morgan (1974). An example of the former application is the work of Dearnaley *et al.* (1975) on the study of corrosion films on type 316 stainless steel. Figure 8.9 shows the backscattering spectrum obtained with 2·9 MeV helium ions from a steel specimen corroded in a dilute solution of NaOH. At 2·9 MeV the ^{16}O (α, α) cross section is 35% greater than Rutherford (Cameron (1953)) and this is sufficient to generate a pronounced "shoulder" on the metal continuum corresponding to the additional oxygen counts. The ratio between anions (oxygen) and cations (metal) in the corrosion film can be assessed to an accuracy of about 10% from such data, and by this means the oxide itself can be identified. If the heights of the shoulders for oxygen and at the metal edge are H_0 and H_m respectively, then the metal to oxygen ration R is given by

$$R = \frac{H_m/Z_m^2}{H_0/Z_0^2} \tag{8.16}$$

(For the case of the rounded shape of the metal edge in Fig. 8.9 it is necessary to take the projected height of the number of cation scattering centres at the metal edge.) "Chemical" analyses carried out in this way may be simplified if the specimen under investigation is compared with standard oxide specimens, such as Cr_2O_3 (Fig. 8.9). For the study of corrosion films, Dearnaley and co-workers used compacted powder specimens as standards, mounted on vitreous carbon discs. Since ion backscattering is not able to distinguish between Fe, Ni and Cr present in the steel, powder mixtures of oxides made to a known composition can be helpful in arriving at the nature of the corrosion product. In this case, a mixture of NiO. Fe_3O_4 gave an identical value for the anion:cation ratio as the corrosion specimen. An additional feature of Fig. 8.9 is that the backscattered yield from the corrosion sample rises sharply with depth. This indicates that a substantial proportion of metal is contributing to scattering very close to the surface, implying that there are regions within the corrosion film which contain "islands" of unoxidised metal, or that the film is thin enough for the ion beam to penetrate through to the underlying metal. Optical examination of sectioned samples showed the second alternative to be the case. The characteristic of showing an increased metal yield with depth was in contrast to the other films examined (corroded in Li-containing solutions) and highlighted the extreme non-uniformity in thickness of the films which would have made examination by more conventional techniques difficult. The rounding off at the metal edge implies also that a considerable proportion of the metal has been replaced by other species (i.e. oxygen). A similar effect has been observed due to surface roughness, Barragan (1974), Schmid and Ryssel (1974). SEM examination of the corrosion specimens typical of Fig. 8.9 showed the surface to consist of numerous crystallites. (The ability to examine as-received samples without the need for elaborate specimen preparation is a great

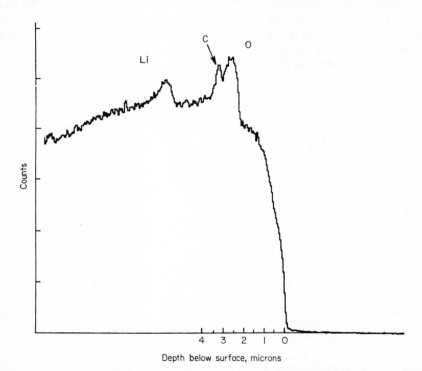

FIG. 8.10. 2·0 MeV H^+ backscattering spectrum from a similar specimen to the previous figure, here corroded in a solution containing LiOH. The greater penetration of protons, compared with He^+ ions, at this energy is evident. The oxygen peak corresponds to complete transversal of the corrosion film and can be used to measure the oxide thickness. Dearnaley et al. (1975).

advantage of Rutherford backscattering in such cases where the nature of the corrosion film can be destroyed by excessive handling.) Figure 8.10 is the backscattered spectrum from a similar specimen, corroded this time in LiOH + NaOH solution. Protons at 2·0 MeV were used to examine the sample so that the oxide thickness could be determined. The depth scale is compressed and the mass resolution is decreased because of the extra penetration of protons and their smaller mass. There is sufficient oxygen present in such samples to enable an estimate of the oxide thickness to be made using equation (8.12). The oxide thickness was found to agree well with subsequent optical measurement. Figure 8.10 illustrates some other features common to backscattering: firstly, the presence of a C peak, arising partly due to cracked hydrocarbons on the surface of the sample during analysis, but due also to carbon taken up in the outermost regions of the corrosion product, presumably during exposure to the solutions; secondly, the presence of lithium, visible now because the beam is able to sample the whole Li distribution while still retaining a substantial proportion of its energy loss (scattering from such light nuclei, however, requires the centre of mass correction, equation (8.8) to be applied). If the carbon peak is subtracted from the oxygen counts, it can be seen that the peak shapes of

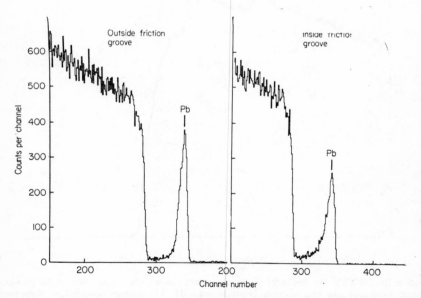

FIG. 8.11. Backscattering data from 2·0 MeV He$^+$ ion microbeam experiments from friction experiments on En 352 steel implanted with 10^{16} ions per cm^2 of Pb$^+$ at 175 keV (Hartley et al., 1973).

oxygen and lithium are approximately similar. This implies that the lithium is uniformly distributed through the corrosion film. (A more quantitative analysis of lithium was undertaken at 3·0 MeV using the ^{7}Li (p, α) α reaction, which is described in Section 8.5.) Finally a characteristic of proton backscattering spectra is the decrease in continuum counts as the energy of the beam decreases (i.e. with greater depth below the surface). This arises because although the cross section is increasing as the energy falls (equation (8.5)), for protons the rate of change of energy loss per unit mass traversed is approximately 4 times lower than that for helium ions at the same energy, and so the net number of counts for a given depth of specimen is small. This is in direct contrast to the case of alpha-particle scattering, where energy loss processes are sufficient to produce a large number of counts at the first few channels resulting in the characteristic upturned shape of such spectra (Fig. 8.11). During backscattering experiments it is frequently necessary to suppress the large number of low energy counts by applying a threshold voltage at the input end of the multi-channel analyser. The effect of this is to increase the sampling efficiency by reducing the dead time, i.e. the time for which the data analysis is effectively accepting no pulse height information.

The particles which arrive at the silicon surface barrier detector are not spatially resolved. Thus a backscattering experiment cannot incorporate variations in lateral density of the implanted species, for example, since all the particles scattered over the beam area falling within a certain solid angle will be detected indiscriminately with respect to their starting position. Backscattering spectra refer to an integrated density distribution, or thickness of corrosion product (Fig. 8.9). The most convenient method of obtaining lateral information is to use ion beams of very small dimensions and to arrange for the sample to be positioned systematically at selected areas of interest. Figure 8.11 shows the backscattering spectra of Pb obtained from within and outside a friction groove 120 μm wide in the surface of an ion implanted steel test piece (Hartley et al., 1973). In this case a 15 μm wide beam of 2 MeV He$^+$ ions was used to bombard the specimen to examine the distribution of Pb after sliding tests. The alteration in peak shape indicates that the lead has become redistributed below the surface (there are more Pb counts in the "saddle" region for the spectrum in the right hand half of Fig. 8.11), and so even after the locally very high adhesive forces encountered during sliding the majority of the implanted Pb ions are still present. By stepping the specimen through very small intervals using a specially constructed target chamber, a similar study of friction grooves, Hartley et al. (1974) showed how implanted species are distributed. The dopant concentrations can then be matched with the mechanical surface profile of the groove in the same region to relate the dopant distribution to surface roughness.

Special refinements to the backscattering technique have included the use of grazing angles of incidence (Section 5.6) to improve the depth resolution (Grant and Williams, 1975) and the use of medium and low energy ions (Buck and Wheatley, 1972, Ball et al., 1972). At energies below 100 keV backscattering is essentially biassed towards investigations of the outermost monolayers of a target, since the ion beam is neutralised rapidly as it undergoes collisions. As the bombarding energy is reduced to a few keV the collision cross section is no longer of the Rutherford form and is therefore no longer given by equation (8.5). Instead, cross sections are derived from the screened Coulomb potential (Everhart et al., 1955), discussed in Chapter 2. The use of elastic scattering of rare gas ions (He^+, Ne^+, A^+) at energies below 10 keV has recently appeared in a potentially powerful method of examining the top monolayer of solid surfaces in what has been termed Ion Scattering Spectrometry (see Section 8.7) Honig and Harrington (1973). Backscattering experiments at MeV energies with heavier ions than helium, for example carbon (Abel et al., 1973 a, b), and nitrogen (Dearnaley and Squires, 1975, Alexander et al., 1973) give increased depth and mass resolution over a narrower region close to the surface.

In conclusion, the conditions for backscattering will normally be defined by the energy and versatility of the accelerator which is used. The optimum choice of ions and energy is a compromise between the high mass and depth resolution obtained at low energy with heavy ions, because of the $(Z_1/E)^2$ dependence of the cross section, and the advantages of deeper penetration which go with the use of low mass ions at high energy (e.g. protons at 2·0 MeV). Backscattering is ideally suited to studying the migration of ion species by interdiffusion or corrosion. In favourable cases several significant features of such phenomena can be gained from the inspection of data after only a few minutes exposure to an ion beam. It may be necessary to expose the sample to one or two additional energies to resolve ambiguous features, but this disadvantage must be weighed against the convenience of a technique which requires neither special specimen preparation nor UHV and examines a depth normally inaccessible by other methods. The extension of backscattering to physical and chemical problems outside the field of semiconductors is a natural progression of its versatility. In the following section we will consider some experimental results for the special case of single crystal targets.

8.3 CHANNELLING

It has been implicitly assumed so far that the target material is amorphous. When the target has a highly ordered structure then its periodicity can under favourable circumstances have a profound effect on the penetration of the beam. The incoming high energy ions may become stably deflected down channels or between crystal planes, suffering greatly reduced energy loss through small angle electronic collisions with the channel walls. The theoretical treatment of channelling has been outlined in Chapter 3, and we will discuss in this section some of the applications and experimental methods relevant to the technique. For semiconductors the topic of channelling has been well covered (e.g. Mayer *et al.*, 1970) since it serves as a convenient method of introducing dopant species up to 10 times the normal range, but it can also perturb intended profiles by causing anomalous "tails" to be produced (Blood *et al.*, 1974). As an analytical tool channelling has found extensive use in the field of semiconductors, and its application to metallurgy

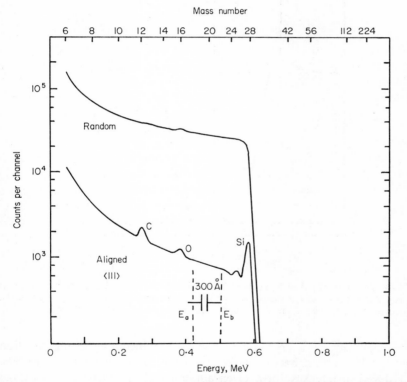

FIG. 8.12. Aligned ⟨111⟩ and random backscattering spectra for a silicon crystal recorded for the same dose of 1·0 MeV He$^+$ ions. The depth resolution and mass scale are included. The more pronounced peaks on the aligned spectrum correspond to small amounts of carbon, oxygen and silicon which are not aligned with the underlying substrate (Davies *et al.*, 1967). The energy range between E_a and E_b represents a region of the spectrum which may be mentioned for setting up channelling experiments. See text and Appendix III, Fig. III.1.

is beginning to be realised more widely. Comprehensive treatments of channelling which have appeared most recently are a review by Gemmell (1974) and the book by Morgan (1973). Clearly, for metallic implantation work channelling will find little application since it is only the most esoteric of studies which requires highly perfect single crystal metal specimens. But for the analysis of the lattice positions of various atoms and fundamental alloy theory the applications of channelling look most promising. These fields have been pioneered by Davies *et al.* (1967), Alexander *et al.* (1970, 1973, 1974), de Waard and Feldman (1974) and Poate *et al.* (1974).

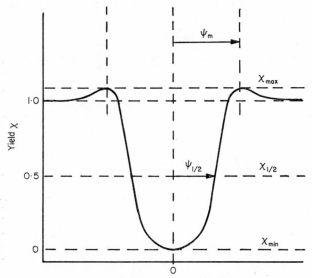

FIG. 8.13. The normalised yield at fixed interaction depth of a close encounter process such as a Rutherford collision measured as a function of angle in channelling experiments. $\chi = 1$ corresponds to the yield from a hypothetical amorphous target (after Gemmell, 1974).

The alignment of a crystal direction to the incoming beam direction causes the Rutherford backscattering spectrum to become altered in shape and reduced in intensity for a given amount of beam charge, because the particles penetrate deep into the substrate and are lost to the detector (Fig. 8.12). To determine if the crystal is near a channelling direction it is necessary to monitor the number of counts arriving at the analyser as a function of time or total beam charge. For ideal crystals with perfect surfaces this could be achieved by simply observing the magnitude of the surface peak (Fig. 8.12) but in practice it is more convenient to select a band of energies between E_a and E_b and watch for a decrease in counts in this region. The "minimum yield" (Fig. 8.13) is a measure of the extent of close collisions and the perfection of the crystal. An electronic discriminator (see Appendix III) is therefore used which is essentially a gate applied to the main amplifier so that only pulses between a specified range are counted. For approximate alignment it is sufficient to feed the output pulses through a linear rate meter with a dial type display, but for more accurate fixing a scaler (counter) is used whose numbered display is independent of small variations in beam current. Having found the channelling direction of interest the experiment can proceed. For example, the changes in the proportion of interstitial (channel-obstructing) sites are annealing the crystal can be determined. Dechannelling (Grasso, 1973) affects the rate of take-up of counts below E_a and is a measure of the local integrity of a crystal. Double alignment refers to the independent alignment of a detector with respect to a second channelling direction (Picraux et al., 1972) and provides a further 30–100 fold attenuation in the observed yield from lattice sites and an approximately threefold increase in yield from midchannel sites. For higher beam energies the channel width becomes narrower since

$$\psi_1 = \left(\frac{2Z_1 Z_2 e^2}{Ed}\right)^{1/2} \tag{8.17}$$

where ψ_1 is the measured halfwidth of the channel (Chapter 3). Under these conditions it is necessary to search for smaller changes in the number of pulses between a selected range, and it is therefore advisable to have a steady beam. Intrinsic to channelling studies, therefore, is the accurate and reproducible remote manipulation of the specimen through small angles. Several goniometer target stage designs are now available and a versatile combination is the three-axis goniometer of Knox (1970) controlled by a McLean stepping motor control unit (e.g. see Section 7.14).

The sensitivity of channelling to perturbations from close interactions with lattice atoms effectively allows atomic displacements as small as approximately 0.1 Å with respect to an aligned set of rows or planes to be recognised. Although in principle any close interaction process (Rutherford scattering, nuclear reaction or X-ray emission) can be monitored in channelling experiments, in practice the majority of studies have involved Rutherford scattering and we restrict ourselves to this case here. Channelling, and its inverse "blocking," have found extensive application in the measurement of short nuclear lifetimes in crystallographic studies using β particles. For a variety of other fields see, for example, specific chapters in the book by Morgan (1973). If the angle of incidence of a beam to a channelling direction is altered slightly, substitutional impurities will produce the same angular dependence for the yield of counts as the host lattice, while for non-substitutional impurities the yield will be different from the host. The application of channelling forms the basis of lattice location experiments, and by arranging for angular scans in three different directions it can readily be seen (Fig. 8.14) that the position of an impurity within the lattice can be determined (Davies, 1973). If, for example, the foreign atom is substitutional, it lies within the shadow of all close-packed rows in the lattice and therefore large attenuations in close impact phenomena with this species will be observed along both the $\langle 01 \rangle$ and the $\langle 11 \rangle$ directions for the two-dimensional "simple cubic" lattice of Fig. 8.14 (Davies, 1973). Atoms positioned interstitially will limit the amount of attenuation (i.e. minimum yield) that is observed, depending on the symmetry of their position with respect to the lattice structure and the channelling direction. In principle, then, it is possible to "triangulate" on impurity atoms using channelling and to determine their lattice positions. However, the situation may be complicated by such factors as the non-linear distribution of the density of channelled ions across the channel ("flux peaking") (see also Chapter 3). The existence of a non-uniform flux for a channelled beam means that the interaction yield for interstitial atoms depends on their position in the particular channel. The significance of flux

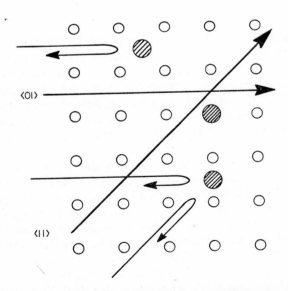

FIG. 8.14. A two-dimensional model of a crystal with impurities showing how the channelling effect may be used to locate lattice sites. The three positions for a foreign atom can be distinguished by comparing the channelling yield along the $\langle 01 \rangle$ and $\langle 11 \rangle$ directions.

peaking is that, unless the implanted ions are entirely substitutional, implanted atom location now requires not only a single measurement of the relative impurity to host backscattering yield along the axis of a particular channel, but the more laborious recording of the yield as a function of angle of impact of the channelled ion beam relative to the crystal axis. An excellent example of the elucidation of quite complex lattice preference sites for Br in Fe making use of such angular scan methods is given by Alexander *et al.* (1973). An additional uncertainty of channelling experiments is the fundamental interpretative problems which still exist (de Waard and Feldman, 1974) regarding the precise proportion of interstitial versus substitutional sites for metallic implantations deduced from minimum yield data. One problem is the preparation of analogue "alloys" on which to conduct channelling experiments. It has been suggested (Abel *et al.*, 1973c) that the vicinity of an implanted impurity may be highly strained, thus distorting the minimum yield. Although accounting for the minimum yield of certain systems is difficult, channelling studies could make significant contributions to models for the short range structure in alloys. The recent work of Poate *et al.* (1974) indicating that 100% substitutionally (i.e. perfect mixing) for room temperature implants of Au in Cu reinforce the conviction that high solubility does imply high substitutionality and that alloy formation can be induced by implantation, for systems where the solubility range extends right across the phase diagram. It is possible that implanted species come to rest in quasi-substitutional positions surrounded by vacancies. Channelling experiments cannot *a priori* distinguish between such apparent locations and their substitutional sites, although vacancies and other radiation-induced lattice disturbances will affect the dechannelling rate. To date lattice location studies on unimplanted alloys, for example by Mössbauer spectroscopy (or hyperfine interactions), are relatively few and restricted to specific hosts such as Fe and Al. The application of channelling to alloys in which stable precipitates form is yet to be established since the coherence or lattice interface matching, across such precipitates may be low resulting in a matrix no longer suitable for channelling experiments.

The minimum impurity concentration that it is possible to investigate by channelling depends on the system and the probing conditions. The data reported have all referred to impurity concentrations of at least 0·1% usually implanted within 1000 Å of the surface. Care has to be taken in such cases to minimise damaging the lattice since MeV probing beams directed at single crystal targets will cause displacements. For metallic targets, in which the damage anneals rapidly the displacement damage has been reported to amount to 5% of the implanted atoms for Fe (de Waard and Feldman, 1974).

These authors, in a study which combined hyperfine interaction studies and channelling, implanted a variety of different species of varying chemical and physical properties into Fe. It was deduced from ⟨111⟩ axial and (100) planar channelling experiments that high substitutional fractions of ions such as I and Xe occur even though their difference in ionic radii is greater than the 15% limit suggested by the Hume-Rothery rule (Hume-Rothery et al., 1969). Figure 8.15 shows the result of ⟨111⟩ angular scans on Fe for the implanted species, all of which show a degree of substitutionality, including Te which is considered insoluble. If the minimum yields for the implanted

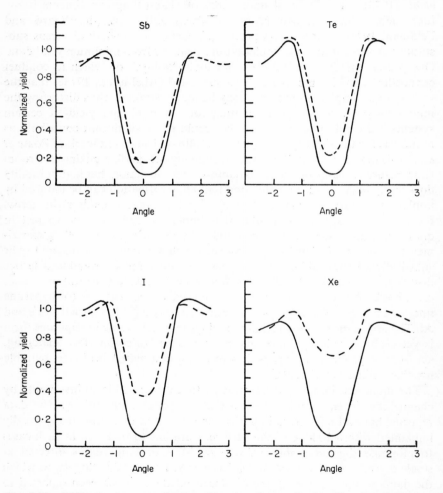

FIG. 8.15 ⟨111⟩ angular scans for iron host (continuous lines) and various impurities (dashed lines) implanted at room temperature to doses between 5·5 and 8 × 10^{14} atoms/ cm^2 at 100 keV. Channelling was carried out with 1·8 or 2·0 MeV He$^+$ ions. For clarity the experimental points are omitted (after de Waard and Feldman, 1974).

and host species are χ_{imp} and χ_{host} respectively, then the substitutional percentage is related by

$$R = \frac{1 - \chi_{imp}}{1 - \chi_{host}}. \qquad (8.18)$$

To date the metallic lattices which have been probed by channelling for impurities location are Fe, W, Nb, Ni, Cu, Al, Au, Pd, Ag and Ti. A list of the positions of a variety of implanted species is given by de Waard and Feldman, who characterised the behaviour as interstitial, substitutional or random according to the value of R in equation (8.18). In a related study the site occupancy of carbon in iron was established, Feldman, 1973, to be $\frac{2}{3}$ on octahedral sites and $\frac{1}{3}$ randomly distributed. In this case the ^{13}C and ^{12}C (d, p) reaction at 1·15 MeV was used. (Nuclear reaction methods are discussed in Section 8.5.) Figure 8.16 is a Darken–Gurry plot (Darken and Gurry, 1953) summarising channelling data for various impurities in Fe (de Waard and Feldman, 1974). This type of plot which shows each impurity as a point on a graph of radius versus electronegativity is used in alloy theory to indicate those elements which would be expected to alloy substitutionally with a given host. The materials of preferential solubility will fall within an ellipse centred on the host, the major and minor axes being defined

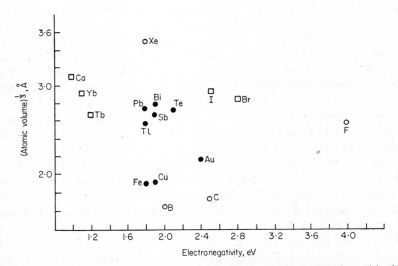

FIG. 8.16. Darken–Gurry plot summarising channelling data for various impurities in Fe; see text for details. The symbols correspond to; ●, 80–100% substitutional; □, 50–80% substitutional; ○, < 50% substitutional. The atomic volume is in arbitrary units and the electronegativity follow Paulings scale (de Waard and Feldman, 1974).

by alloying rules such as those of Hume-Rothery. It is interesting to note that Tl, for example is in excess of the 15% variation in radii which is one of the normal alloy criteria (i.e. for ions outside of this size range it is considered unlikely that a given binary mixture will combine substitutionally), although channelling and hyperfine interaction studies both indicate a very high degree of substitutionality for this element. Considerations of the role of local vacancies or stress-induced perturbations of the type discussed by Sood (1975) may lead to further understanding of how certain species can occur with the minimum free energy position at a lattice site when geometrical factors indicate this to be unfavourable.

In this section we have dealt almost exclusively with lattice location applications of channelling since for the metallurgist this is the most relevant area. In conclusion reference should be made to channelling studies on metal oxides Matzke *et al.* (1971), Della Mea *et al.* (1975) and silicon nitrides (Gyulai *et al.*, 1970) which show by the methods indicated above, how the oxygen or nitrogen atoms are incorporated in the lattice. Energy loss measurements may also be performed in such systems since, for the case of TiO_2, open hexagonal channels exist resulting in well characterised channelled beams (Della Mea *et al.*, 1975). When combined with computational models incorporating the effects of flux peaking and interatomic potentials as discussed in Chapter 3, channelling is an extremely powerful method of probing the atomic environment of crystalline systems. In addition to metal samples, alkali halides and oxides have been studied to determine the extent of channelling in well-characterised ionic lattices (Whitton and Matzke, 1966), and also as part of a larger investigation on the channelled energy loss in transmission experiments on materials covering a range of Z_2 values (Clark *et al.*, 1970). In channelling experiments with 40 keV Xe^+ ions in single crystals of NaCl, KBr, MgO, SiO_2 and UO_2 Whitton and Matzke found the effect of bombardment dose on the range profiles to resemble the behaviour of semiconducting materials, i.e. with damage by displacements towards the end of the channelled trajectory. The dose required to cause a crystalline to amorphous transition was found to be at least an order of magnitude lower in MgO and SiO_2 which are known to be damaged relatively easily by ion bombardment (Matzke, 1966). Clark *et al.* (1970) took advantage of the ready availability of alkali halide crystals (NaCl, CsI, also MgO) to investigate the effects of channel width on energy loss. In agreement with the work of Sattler *et al.* (1967, 1968) on semi-conductor targets these authors found that there was no significant dependence of the peak width upon the electron orbital configuration and the dominant factor was shown to be the width of the channel, although the relative contribution to peak width of straggling effects is still open to question (Appleton *et al.*, 1967) as discussed in Chapter 2.

8.4 RESONANT ELASTIC SCATTERING

In Section 8.2 we considered purely elastic events of Rutherford scattering in which the electronic charge of the projectile and target atom are mutually repelled by Coulombic interaction, and the scattering cross section is readily calculated using equation (8.6). Reference was also made to an energy regime for which the cross section for oxygen is 35% higher than the Rutherford case, making the detection of oxygen in corrosion films relatively

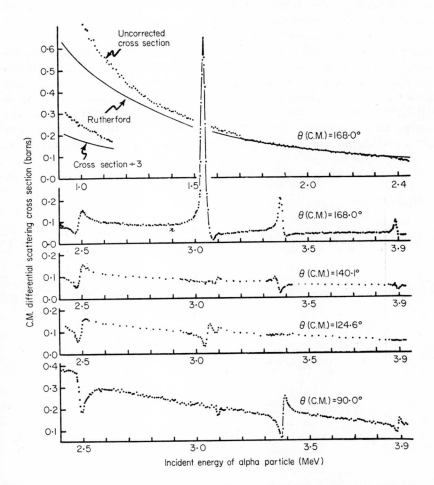

FIG. 8.17. The ^{16}O (α, α) ^{16}O experimental cross sections at 168·0° in the region of 0·94 to 2·4 MeV and also at 168·0°, 140·1°, 124·6° and 90° in the 2·4 to 4·0 MeV energy range (Cameron, 1953).

straightforward (Fig. 8.9). In this section we discuss resonant elastic scattering in greater detail.

The excitation curve of a particular nuclide, that is the scattering cross section or particle yield for a reaction, is usually of complex shape above about 2·0 MeV as the nucleus itself becomes penetrated by the incident ion. The result of such close encounter processes is that various resonances and energy level transitions are induced. If a particular excited state is favoured then the capture cross section of the target atom will be enhanced at a certain projectile energy. The shape of such excitation curves can in many cases be predicted by theoretical treatments of the spin and angular momentum properties of the nucleus. For the analyst it is useful to be aware of non-Rutherford type variations in cross section since deviations may be troublesome, or alternatively, used with advantage. Figure 8.17 shows the variation in scattering cross section for helium ions incident on ^{16}O at various scattering angles and is taken from the original paper of Cameron (1953). Such graphical representations of the cross section are essential in order to select the energy most suitable to excite either a broad or narrow resonance. For accurate determination of metal/oxygen ratios and oxide thickness it is

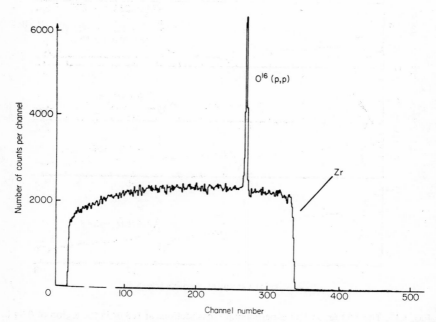

FIG. 8.18. Proton elastic resonant backscattering from an oxidised zirconium sample (Wiedman, 1973).

necessary to choose regions over which the cross section is either smoothly varying or constant. For helium ion bombardment two such energies are 1·8–2·2 MeV where the cross section obeys the Rutherford law, and 2·6–2·9 MeV where σ remains constant for detection at around 165° to the incident beam. In this way, the oxygen present in the corrosion film of Fig. 8.9 was detected (Dearnaley et al., 1975).

Although in resonant elastic scattering the abnormal yield is a nuclear effect, it is not a nuclear reaction. Nuclear reaction techniques, in which the particle detected is different from the incident particle and the scattering is inelastic, are described in the following section. For the backscattering experiments already discussed, the use of elastic resonances merely enhances the detection of specific elements. The scattering event is still elastic and the energy of the scattered particle is predictable from equations (8.1) and (8.3). A second example of the use of elastic resonances also concerns a study of oxidation. The conventional method of establishing oxygen uptake as a function of time is by weight gain but this can be inaccurate from the small

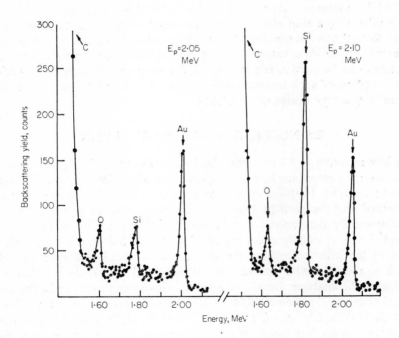

FIG. 8.19. Backscattering of protons from a carbon substrate coated with very thin layers of silicon oxide and Au. At a proton energy of 2·05 MeV the yield ratio of Si to Au is about $\frac{1}{2}$, and at $E_p = 2·10$ MeV the yield ratio is 2·0 due to $^{28}Si(p, p)\,^{28}Si$ resonance at the higher energy (Chu et al., 1973).

specimens typically used for implantation. Dearnaley et al. (1972) have used the very large elastic scattering cross section of ^{16}O for protons above 4 MeV. Over a wide range of scattering angle and energy the differential scattering cross section remains constant and is of the order of 80 times the Rutherford value (Laubenstein et al., 1951). Figure 8.18 illustrates the detection of oxygen on implanted zirconium examined with protons at 4·55 MeV (Wiedman, 1973). This resonance region was also employed by Rickards and Dearnaley (1974) to determine the oxygen uptake on implanted copper. An alternative resonance which occurs at a lower energy with helium is the one at 3·05 MeV (Figure 8.17). Elastic (p, p) and (α, α) resonances have been observed for many light nuclei at various energies. For example ^{28}Si$(p, p)^{28}$Si was used to study the energy levels of ^{28}P(Verona et al., 1959), and ^{12}C$(\alpha, \alpha)^{12}$C, ^{14}N$(p, p)^{14}$N and ^{16}O $(\alpha, \alpha)^{16}$O for the study of ^{16}O, ^{15}O and ^{20}N levels (Hill, 1953, Bashkin et al., 1959). The energy levels of light nuclei are tabulated by Ajzenberg-Selove and Lauritsen (1968, 1974) and will be described further in Section 8.5. Chu et al. (1973) include a discussion of resonant elastic scattering reactions and have used protons at 2·10 MeV to demonstrate the effect of a resonant proton reaction with ^{28}Si by bombarding a thin film specimen mounted on a carbon substrate (Fig. 8.19). With the use of standard specimens to avoid the calculations necessary to arrive at absolute values of oxygen on silicon, for example, resonant reactions can be particularly useful for the detection of such light species. The adoption of such methods will be restricted, however, to laboratories where high energy beams are available.

8.5 NUCLEAR REACTION ANALYSIS

The low sensitivity of Rutherford backscattering for light elements can in many cases be overcome by the complementary detection method of nuclear reaction analysis. In this case the scattering is inelastic, kinetic energy is not conserved and the particle detected is different from the incident particle. Experimentally the procedure is identical to that used for backscattering, except that it is usually necessary to place a thin foil (e.g. 0·0001" Al) in front of the detector to limit the amount of elastically scattered particles which would otherwise saturate the detector and contribute to pile-up of counts on the analyser. Details of nuclear reaction methods can be found in Amsel et al. (1971), Pierce (1974) and in the more general review articles of Wolicki (1972), Chu et al. (1973) and Bird et al. (1974). The two basic advantages of nuclear reactions with respect to backscattering are that they provide a background free detection of light elements on heavy substrates and that they allow ideal discrimination between two isotopes of the same element. The latter property has been used extensively in recent studies of the oxidation mechanisms of various metals and alloys taking advantage of the ^{18}O (p, α) ^{15}N reaction (see review articles and references below). The background free detection is due to the high positive Q values of many nuclear reactions which result in the emission of particles with energies well above that of the beam. A nuclear reaction spectrum may therefore consist of a series of peaks each of which relates to the yield of a particular reaction induced in the target. The peak shape depends on the cross section and resonance width of the particular reaction as well as the beam energy with respect to the nuclide position below the surface. Nuclear reactions behave in an opposite manner to backscattering where heavy nuclei on light substrates are easiest to detect. A combination of both methods, using essentially the same equipment, is therefore a powerful analytical tool (Abel *et al.*, 1973a, b; Dearnaley *et al.* 1975).

A typical nuclear reaction may be written as ^{18}O$(p, \alpha)^{15}$N, or alternatively

$$^{18}_{8}\text{O} + ^{1}_{1}\text{H} \rightarrow ^{19}_{9}\text{F}^* \rightarrow ^{15}_{7}\text{N} + ^{4}_{2}\text{He} + \gamma + Q \tag{8.19}$$

where Q is the energy released during the reaction. With reference to the symbols defined in Appendix III the conservation of mass and energy requires that the total energy E_T is given by

$$E_1 + Q = E_3 + E_4 + E_\gamma - E_T \tag{8.20}$$

and

$$Q = (M_1 + M_2 - M_3 - M_4) c^2. \tag{8.21}$$

Momentum is also conserved so that

$$M_3 V_3 \cos \theta + M_4 V_4 \cos \phi = M_1 V_1$$
$$M_3 V_3 \sin \theta + M_4 V_4 \sin \phi = 0. \tag{8.22}$$

These equations determine the product particle energies and their dependence on incident particle energy, reaction Q-value and angle of observation (Bird *et al.*, 1974). The larger the value of Q the higher will be the energy at which the product particle emerges from the close encounter. Q-values are both positive and negative (in which case there is a threshold energy below which the reaction cannot occur). The first stage in the process is for the Coulomb barrier to be overcome as discussed in the previous section. Generally, this will occur (Chapter 2) for a target-projectile pair for incident energies $E_{1_{CB}}$ where

$$E_{1_{CB}} \geq \frac{Z_1 Z_2}{(M_1^{1/3} + M_2^{1/3})} \text{ MeV}. \tag{8.23}$$

The Q-value of the reaction will define whether or not reaction products can emerge. The detection of a target species depends also on the capture cross section. Provided the cross section is large, detection of a particular species may be achieved by using a strong, broad resonance (e.g. $^{11}B(p, \alpha)^8Be$ at 1·4 MeV) in which case depth information may be difficult to obtain accurately. Alternatively, a narrow resonance (e.g. $^{18}O(p, \alpha)^{15}N$ at 1·763 MeV, width 4·0 keV) may be used to determine concentration profiles by exciting the resonance at specific depths below the surface and calculating the appropriate scattering centres at that depth (Neild *et al.*, 1972; Barnes *et al.*, 1973).

Q-values may be referred to in Nuclear Data Tables (Wapstra and Gove, 1971) and the reactions for light nuclei are discussed by Ajzenberg-Selove and Lauritsen (1968) where energy level diagrams, together with sketched forms of cross section variations, are given. If the target atom is promoted to an excited state during the reaction, then the energy of the emitted particle is reduced by an amount of energy equal to the excitation energy. The existence of a set of excited states leads to the observation of a spectrum of discrete

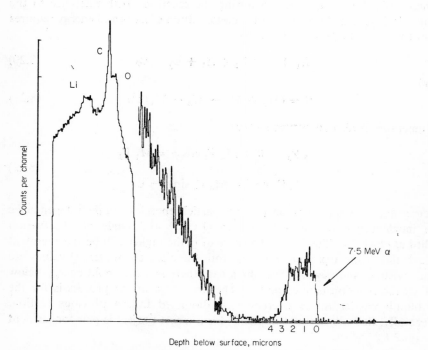

FIG. 8.20. 3·0 MeV proton backscattering spectrum from LiOH-corroded type 316 stainless steel. The higher energy peak (shown on a magnified vertical scale) is from the $^7Li (p, \alpha) \alpha$ nuclear reaction.

product particle energies (and also of gamma ray energies detected using NaI crystal detectors) which are characteristic of the final nucleus. In equation (8.19) ^{18}O is promoted for a short time (10^{-14} s) to ^{19}F which then decays to ^{15}N releasing an α particle and a γ ray. There is thus an advantage of nuclear reaction analysis over Rutherford backscattering in that it is specific to particular isotopes. Energy level diagrams are complicated and it requires a certain familiarity with the methods of nuclear physics to select a suitable detection angle and energy of incident ion (optimised according to the functional dependence of the scattering cross section) such that the yield from competing reactions does not obscure the data of interest. Fortunately for the analyst many review articles list suitable reaction energies and detection rates, although it is frequently helpful for more detailed treatments of a particular reaction to consult the original reference. Table 8.1 (Chu *et al.*, 1973) lists such parameters for detection experiments.

TABLE 8.1 Typical experimental conditions for nuclear microanalysis (Chu *et al.*, 1973).

Nucleus	Reaction	Bombarding energy (keV)	Energy of emitted particles (MeV)	Thickness of Mylar absorber (μm)	Counting rate[a]
^{16}O	^{16}O$(d,p)^{17}$O*	830	1·52	19	1600
^{18}O	^{18}O$(p,\alpha)^{15}$N	730	3·38	12	4600
^{14}N	^{14}N$(d,\alpha)^{12}$C*	1300	6·76	19	470
^{19}F	^{19}F$(p,\alpha)^{16}$O	1260	6·93	26	265
	^{19}F$(p,\alpha)^{16}$O	1340[c]	6·97	31	1410
	^{19}F$(p,\alpha\gamma)^{16}$O	870[c]	7·12, 6·92, 6·13 $\}\gamma$	—	13800[b]
^{12}C	^{12}C$(d,p)^{13}$C	1000	3·01	19	20000
^{2}H	D(d,p)T	550	2·45	6	2000
^{6}Li	^{6}Li$(d,\alpha)^{4}$He	1560	9·25	26	930
^{7}Li	^{7}Li$(p,\alpha)^{4}$He	1000	7·84	19	400
^{27}Al	^{27}Al$(p,\gamma)^{28}$Si	992	—	—	—
B	^{11}B$(p,\alpha)^{8}$Be	700	4·07	—	—
B	^{10}B$(n,\alpha)^{7}$Li	Thermal neutron	1·47 and 1.78	—	—

[a] for a film containing 10^{16} atom cm^{-2}, a 1μA beam, per min. The solid angle is 0·12 steradian, at 150° detection angle. The counting rates may be multiplied by 3, using 3 identical detectors, when necessary.
[b] 3 in. × 1 in. NaI(Tl) scintillation detector at 7 cm.
[c] for films of equivalent thickness < 10 keV.

Figure 8.20 shows the spectrum obtained from the bombardment of a corrosion film, similar to Fig. 8.2, which contains lithium. In this case the lithium concentration is revealed on an expanded vertical scale, by the high energy α counts induced by the ^{7}Li$(p, \alpha)\alpha$ reaction at 3·0 MeV. Once again, the simplest method of arriving at the concentration of a particular species from nuclear reaction studies is to carry out comparison experiments with standard samples. In the work referred to here (Dearnaley et al., 1975) a single crystal of LiNbO$_3$ was used for this purpose. The lithium to metal content of the corrosion film was calculated to be approximately 0·23:1 and when combined with backscattering data and EDAX (energy dispersive X-ray analysis) the corrosion film was deduced to be LiFeO$_2$.NiO, Fe$_3$O$_4$. If standard comparison specimens are not used it is necessary to calculate the yield from a particular reaction as a function of depth (i.e. decreasing beam energy) and this may be complicated. Amsel et al. (1971) discuss the application to thick targets. The interpretation of spectra from resonance reactions excited below the surface depends on the straggling in the absorber

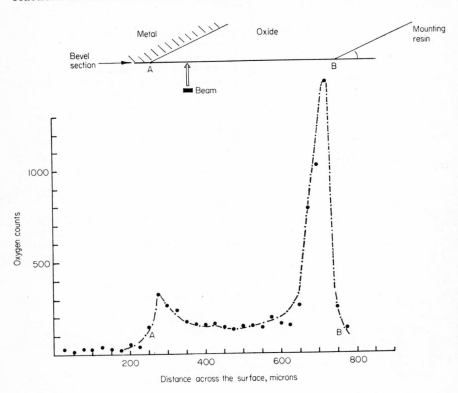

FIG. 8.21. Concentration of oxygen across a 25° bevelled section of an enriched-oxygen corrosion film on 9% Cr steel, obtained for a total beam charge of 10μC using the ^{18}O $(p, \alpha)^{15}$N reaction at an incident proton energy of 840 keV. The beam width was 23 microns, (Allen et al., 1975).

films placed in front of the detector and also on the smaller dE/dx associated with the passage of MeV protons or alpha particles through the target. Amsel et al. (1971) used the narrow resonance (width 2·6 keV) at 629 keV for $^{18}O(p, \alpha)^{15}N$ to study oxygen transport during anodic oxidation of tantalum. Calvert et al. (1974) have used the higher narrow resonance for the same reaction at 1·763 MeV to look at oxygen diffusion in steels and have developed a deconvolution technique to extract the original profile (Neild et al., 1972). Their analysis has proved successful in measuring oxide thickness on chromium and oxygen diffusion profiles in TiO_2 and Cr_2O_3. One way of avoiding the complications associated with thick target analysis is to use bevelled sections and scan across the surface with an extremely narrow beam. The yield from a reaction at the surface (or just below it to avoid surface contamination) then relates simply to the corresponding position deep within the specimen. Figure 8.21 is a plot of the concentration of ^{18}O across a corroded steel specimen obtained on the Harwell microbeam facility (Cookson and Pilling, 1970) by Allen et al. (1975).

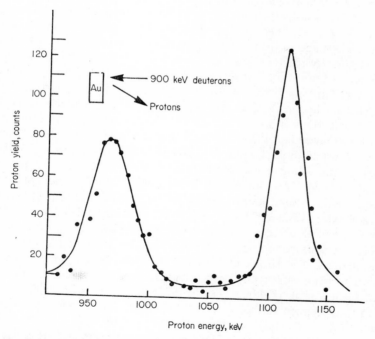

FIG. 8.22. Energy spectrum of protons produced from the $^{16}O(d, p)^{17}O$ nuclear reaction. The two separated peaks show the presence of $(3·4 \pm 0·2) \times 10^{15}$ atoms cm^{-2} of oxygen at the two surfaces of a self-supporting gold single crystal 8000 Å thick (Agius et al., 1975).

The detection limits of light nuclei depend on the particular reaction but in favourable cases such as most of the reactions listed in Table 8.1, 10^{13} atoms cm^{-2} at or near the surface may be detected. This corresponds to about a hundredth of a monolayer. For elements heavier than fluorine the cross sections generally drop by at least an order of magnitude; for example the (p, α) reaction on ^{27}Al and ^{31}P require 10^{15}–10^{16} atoms cm^{-2} in order to obtain reasonable statistics. Agius et al. (1975) in a study of the diffusion of about one monolayer of oxygen through a thin gold crystal established the presence of surface concentrations of oxygen using the ^{16}O$(d, p)^{17}$O reaction at 0·9 MeV (Fig. 8.22).

Other elements which have been examined by nuclear reaction analysis include fluorine, using the ^{19}F$(p, \alpha\gamma)^{16}$O reaction (Möller and Starfelt, 1967). By using the 1·375 MeV resonance in the fluorine reaction Möller and Starfelt were able to analyse a variety of Zircaloy specimens over a depth of two microns and compare the sensitivity of detection with other samples with surfaces prepared by chemical means. Conservative estimates of the sensitivity showed that less than 0·01 µg cm^{-2} of fluorine could be detected using this reaction. An alternative method of detecting fluorine using the ^{19}F$(p, \alpha)^{16}$O reaction at 2·65 MeV, (Case et al., 1975). In this work corrosion films of a few microns thickness were probed to establish the composition profiles of fluorine, carbon and oxygen using the ^{12}C $(d, p)^{13}$C reaction at 1·4 MeV and the ^{18}O $(p, \alpha)^{15}$N reaction described earlier. The ^{11}B$(p, \alpha)^{8}$Be reaction has been employed for boron detection, mainly for channelling experiments (North and Gibson, 1970; Akasaka and Horie, 1973) on ion implanted silicon. Recent applications of this reaction have included the analysis of boron in steels (Olivier and Pierce, 1974). The amount of boron detected in the work of Olivier and Pierce was of the order 1–100 ppm. The lattice site location of carbon in steel, referred to earlier in Section 8.3, was achieved by combining channelling techniques with the ^{12}C$(d, p_0)^{13}$C and ^{13}C$(d, p_0)^{14}$N reactions at 1·1 MeV (Feldman et al., 1973). (The 0 subscript refers to protons emitted from the ground state of the excited nucleus.) Pierce et al. (1974) have also made extensive use of the ^{12}C$(d, p)^{13}$C reaction at 1·3 MeV deuteron energy to examine carbon diffusion profiles in steel. Great care is necessary to avoid spurious counts from hydrocarbon contaminant films deposited during analysis. The effects of contamination can be minimised either by implanting carbon sufficiently deep to ensure clear separation of the counts from a surface peak (Feldman et al., 1973) or the use of a cold trap very close to the specimen surface (Pierce et al., 1974). The sensitivity to carbon of the ^{12}C$(d, p)^{13}$C analytical method was found to be 75–100 ppm for the steels examined by Pierce. The Harwell microprobe has also been used for nitrogen detection (Olivier and Pierce, 1974, Allen et al., 1975). In these experiments the distribution of nitrogen in metal samples over the range

0.05%–2% was determined using $^{14}N(d, p)^{15}N$ and $^{14}N(d, \alpha)^{13}C$ reactions between 1·0 and 2·0 MeV. For deuteron-induced reactions, the Q-values are frequently very large (e.g. $^{14}N(d, p)^{15}N$; $Q = 8·61$ MeV) with the result that the emitted protons, for example, are well separated from the elastic low energy regime of the spectrum and the energy loss through absorber foils is small.

Most of the collections of nuclear reaction data quote Q-values and resonant energies for nuclei above ^6Li. There are a number of reactions, however, which allow the distribution and location of such species as hydrogen, deuterium and helium to be studied. These elements are not easily measured by chemical or mass spectrometry methods (see Section 8.7). Thus hydrogen is often analysed by the use of its isotopes tritium and deuterium. Picraux and Vook (1974) carried out an elegant channelling lattice location study of D and ^3He in tungsten. Ligeon and Guivarc'h (1974) have used the alternative method of inverting the $^{11}B(p, \alpha)^8Be$ reaction to detect hydrogen itself (rather than its isotopes) implanted into silicon by means of the $^1H(^{11}B, \alpha)^8Be$ reaction.

The $D(d, p)T$ reaction has the distinction of having a particularly high Q-value (18·35 MeV) and a relatively large cross section (~ 60 mb/sr) for a deuteron energy around 500 keV. Very few ion beam analyses on the behaviour of hydrogen and helium in solids have been reported, and in view of the vast differences in the quoted values for diffusion coefficients and solubilities for these elements in metals, it is important to establish their physical behaviour more thoroughly. Apart from the general relevance of hydrogen detection methods to stress-corrosion cracking and hydrogen embrittlement, an important topical justification for studies on hydrogen and its isotopes is in power generation. The first wall structural material for controlled thermonuclear (fusion) reactors and the fuel cladding for fast breeder reactors will include transition metals that will be subject to large doses of hydrogen and helium (Picraux and Vook, 1974). In tungsten the He defect configuration is associated with a vacancy, a finding which agrees well with the theoretical calculations of Bisson and Wilson (1974). Deuterium was found to be situated in the tetrahedral interstitial site in W using angular scans along the $\langle 100 \rangle$ direction and $\{100\}$ and $\{110\}$ planes for the $D(d, p)T$ reaction yield in channelling. Picraux and Vook thus established that structural information can be obtained for dilute hydrogen and helium populations in refractory metals using reaction methods. Information on the site location and defect cluster morphology of such species can then be applied to diffusion and solubility models. It was also shown from the sensitivity of the method to damage from the probing beam during analysis or subsequent re-implantation that the early stages of gas bubble nucleation in such materials might be studied by nuclear reactions, since surface gas bubble nucleation reduces the D-induced flux peaking by trapping deuterium on He-rich sites. (Gas bubbles in Mo and Nb are described in Erents and McCracken (1973), and Carstanjen and Sizmann (1972).)

The method of Ligeon and Guivarc'h (1974) for the detection of hydrogen makes use of the symmetry of nuclear reactions. In their case a ^{11}B beam was directed at a hydrogen-implanted silicon target and the reaction yield for α's at a detection angle of 90° with respect to the incoming beam was sufficient to detect surface layers of the order of 5×10^{13} atoms of hydrogen per cm². The profile of implanted H over a depth of 4000 Å with a depth resolution of about 400 Å was achieved. With a boron-11 beam, the equivalent energy to the 163 keV proton-alpha resonance (Huus and Day, 1953) is at 11 times this energy, i.e. 1·793 MeV, and provided experimental facilities exist to handle such high mass beams the potential of such an analytical method for routine hydrogen examination is very large. An obvious extension of the $^1H(^{11}B, \alpha)^8Be$ reaction analysis near 2 MeV is the study of H in metals. Preliminary experiments with a microfocus system at Harwell shows the detection of hydrogen by this means is favourable (Cookson, 1975).

8.6 ION INDUCED X-RAYS

In addition to ejection by sputtering, nuclear reaction or backscattering of an atom from a solid surface bombarded by an ion beam, the collision process between two atoms can result in the emission of radiation. A rearrangement of the atomic orbitals occurs during the overlap of the ion and target atom which leads to inner shell ionisation. The radiation emitted depends on the energy level structure of the target atom, and with electron bombardment the readjustment can either lead to the ejection of Auger electrons (see Section 8.7) or the generation of a series of characteristic X-rays. Electrons induce a continuous spectrum of radiation to be emitted from the specimen, as characteristic X-rays resulting from well defined transitions within the target atoms. Since the end of the last century the standard procedure for generating X-rays has been to bombard a metal target with an electron beam and this still forms the basis for X-ray diffraction crystallography equipment. In this section we shall consider the characteristic X-ray production from bombardment with ion beams—either protons of high energy (MeV) or heavy ions of a few keV energy. The field of X-ray emission spectroscopy is very large and the reader is referred to the review articles of Valković (1973) for a recent appraisal and fundamental treatment of the subject. There are also good introductory chapters in the book by Kane and Larrabee (Gilfrich, 1974) and in the conference proceedings edited by Shinoda et al. (Hink, 1972). The most distinct advantage of using ion beams rather than electrons for X-ray production is that the bremsstrahlung (braking radiation) is orders of magnitude lower. The continuous spectrum arises when energetic electrons undergo step-by-step deceleration in the target. No continuum is generated by photon excitation, and the quantity of continuum generated by particles heavier than electrons is smaller than that generated by electrons by approximately the square of their mass ratios. Proton induced X-ray methods and applications are discussed by Duggan *et al.* (1972) and Folkmann *et al.* (1974). Cairns *et al.* (1973) assess the technique of heavy ion induced X-rays and provide many earlier references to theoretical models and applications.

In common with nuclear reaction analysis, X-ray spectroscopy (or fluorescence) is able to fingerprint individual atomic species by virtue of the

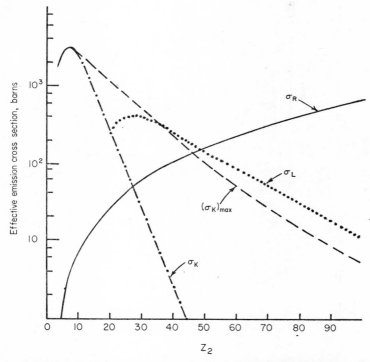

FIG. 8.23. The Rutherford cross section (σ_R) for 170° backscattering of 1·0 MeV He$^+$ ions compared with the effective X-ray emission cross section as a function of atomic number Z_2 of target atoms. The X-ray values are the product of ionisation cross section and fluorescence yield for 2·0 MeV protons. The broken line gives the maximum emission cross section that can be achieved for K X-ray production. This occurs at increasingly higher energies of the incident particle for high mass impurities. The decrease in cross section for low Z elements arises from corrections for absorption in the detection window (Mayer and Turos, 1973).

atomic transitions that occur specific to a particular arrangement of orbitals. (Nuclear reactions, of course, rely on the rearrangement of atomic energy levels within the nucleus and the cooperative effects, or other wise, of angular momenta, spin etc.) This is in contrast with the less specific sampling mechanism of kinematic collisions employed in Rutherford backscattering. One essential difference between Rutherford backscattering and ion induced X-ray analysis is that the X-ray cross section rises dramatically for the light

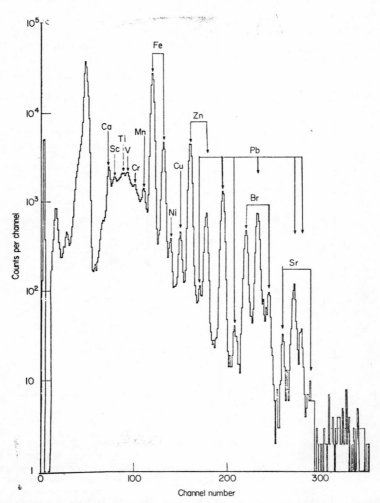

FIG. 8.24. Proton induced X-ray analysis (PIXA) of pollution species trapped in filter paper (Cookson, 1975).

elements, rendering the two techniques essentially complementary (Fig. 8.23). Several processes might result in the removal of an electron from the inner shell in an atom. But we shall mainly be concerned with scattering or excitation processes in which the incoming charged particle (electron, proton, α-particle or heavy ion) collides with an electron in the K-shell and transfers part of its kinetic energy to the electron. As a consequence, a vacancy is created which is filled within $\sim 10^{-15}$ s by an electron from one of the outer shells. If the vacancy is filled with an electron from the L-shell a K_α X-ray will result during this transition (Valković, 1973).

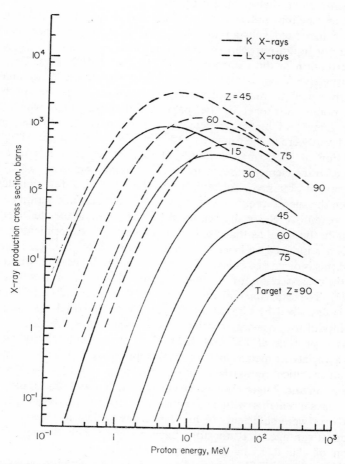

FIG. 8.25. Calculated cross sections for X-ray production by protons (Gordon and Kraner, 1971).

In general X-ray analysis is most suited to the examination of thin targets since the theoretical prediction of X-ray yields for an incident particle of variable energy is complex (Valković, 1973; Hink, 1972). However, with heavy ions the X-ray yield is confined to a narrow region near the surface since the ability of an incident heavy ion to generate X-rays from the target decreases dramatically as slowing down causes its energy to fall towards a cut-off value, described below. Folkmann et al. (1974) have analysed on a theoretical basis the use of protons for X-ray emission analysis and considered the reduction in bremsstrahlung yield from "knocked-on" electrons. They calculate the detection sensitivity obtainable for the concentration of a trace element is to better than 0·1 ppm.

Proton and helium induced X-ray analytical techniques (Feldman et al., 1973) have increased substantially in recent years, largely because of the availability of liquid-nitrogen cooled lithium-drifted silicon detectors which have a resolution in the region of 150 eV for $\sim$ 2 keV X-rays. This makes energy dispersive X-ray analysis feasible. Proton induced X-ray analysis is particularly suited to multielement targets. Figure 8.24 is the spectrum of air-borne pollution trapped in filter paper for an arrangement which favours the measurement of higher Z elements (Cookson, 1975). In this case the impurity levels were in the range 1 to 100 ppm. With thin targets (20 µg/cm^2) the detection of such small amounts of material corresponds to absolute amounts of trace elements down to 10^{-12} g (Johansson et al., 1970). X-ray spectra may be obtained very rapidly—in the order of a few minutes—and apart from the special requirements of a cooled detector and a thin beryllium window to remove elastically scattered particles, the associated electronics is essentially the same as that used for backscattering experiments. (For the "soft" X-rays induced by heavy ions a gas proportional counter is normally used, and this is discussed below.) Protons are normally used at MeV energies because the cross-section for X-ray emission is a maximum at about 2 MeV (Fig. 8.25). The experimental measurement of proton-induced X-ray cross sections is described by Gray et al. (1973). However largely to examine the applicability of lower energy accelerators to X-ray work, Musket and Bauer (1973) used protons at 150 and 350 keV to study thin (20–3000 Å) oxide films on a variety of metals such as 309S stainless steel. Their approach was to make an analytical comparison between proton induced X-rays, Rutherford backscattering and Auger electron spectroscopy. It was found that for oxide thickness measurements with a sensitivity of 10^{-3} monolayers proton induced X-rays were sufficient. In another study which combined the depth resolution advantage of Rutherford backscattering with the superior mass resolution of characteristic X-ray analysis, Feldman et al. (1973) studied anodically grown oxide films on GaAs with He ions, using ion backscattering and He-induced X-rays. These authors were able to conclude from their analysis of such films that when grown under constant voltage the outer surface layer is a pure gallium oxide of $\leqslant 200$Å while the bulk of the film contains Ga:As:O in a ratio of approximately 1:1:2. Annealing to 650°C transforms the film to almost pure GaO. Normally, this type of information would be difficult to arrive at from backscattering experiments alone since gallium and arsenic are comparable in mass. Chemin *et al.* (1973) have applied 0·5 MeV proton induced X-rays to a channelling and lattice location study of P and S in Ge.

In the past few years, strong experimental interest has been directed towards X-rays produced by heavy ions. For example Der *et al.* (1968) measured the carbon K X-ray yield for the ions H^+, He^+, C^+, O^+, Ne^+, Kr^+ and Xe^+ incident on a carbon target at energies from 20 to 80 keV. The cross

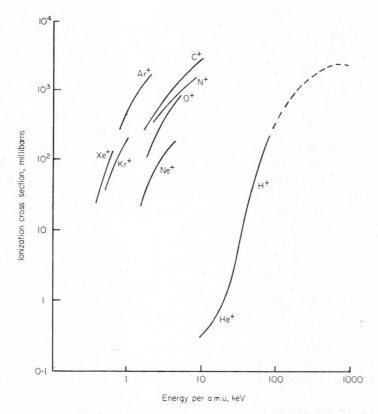

FIG. 8.26. Carbon α-shell ionisation cross section σ_I as a function of incident ion energy per a.m.u. measured by Der *et al.* (1968). For clarity the data points are omitted. The dashed curve represents data from Kahn *et al.* (1965).

sections for the heavy ions were found to be several orders of magnitude larger than those predicted by direct scattering theory as used to describe X-ray production by proton impact (Garcia, 1970, 1973). Figure 8.26 shows the carbon K shell ionisation cross section σ_I as a function of incident ion energy per a.m.u. measured by Der et al. (1968). The dashed curve represents data from Khan et al. (1965, 1966) and it can be seen from the plots that the X-ray production cross sections for heavy ions are even greater than those for protons of the same energy. Later experiments (Fortner, 1969) extended the energy range to 1·5 MeV and the results correlate well with theoretical suggestions that the anomalous X-ray yield is due to an electron promotion mechanism resulting from the interpenetration of the atomic shells of the incident ion and target atom, as originally discussed by Fano and Lichten (1965) for Ar^+–Ar collisions (Saris, 1972; Rudd 1972). Their theory uses a molecular orbital approach in that it considers both the incident projectile and target atom to form a pseudo-molecule at the distance of closest approach. The theory then considers the most favourable amalgamation of orbitals to provide a mechanism for transferring electrons from inner levels, thereby creating vacancies which appear around either the target or the projectile or both. The result is the subsequent emission of characteristic X-rays. A particularly powerful feature of the pseudo-molecule model is that it allows us to understand why certain X-rays arise above a well-defined cut off energy, which corresponds to the distance of closest approach at which an electron can be promoted to a particular shell (Cairns et al., 1973). The ability to excite a soft X-ray component of a particular element by carefully choosing the impact energy makes heavy ion induced X-ray analysis selective to a single element close to the surface. This occurs not solely because of the fall in cross section with energy loss of the incoming ion, but also because of the high absorption of the matrix for soft X-rays. Furthermore, the maxima in cross sections which occur when the target and projectile have matching characteristic X-rays may be explained in the theoretical model of orbital overlap, although the theory does not allow accurate predictions to be made of X-ray cross sections. The physics invoked in the production of inner shell vacancies by proton and heavy ion irradiation is discussed in recent reviews by Garcia et al. (1973) and Kessel and Fastrup (1973).

For the detection of soft X-rays (i.e. within the energy range from 100 eV to several keV) the gas flow proportional counter is the most widely used instrument. The counter usually takes the form of a cylinder with a very thin (typically 10^{-4} cm) plastic window mounted along its diameter, and held at a positive potential of up to 2 keV. The ionisation gas is 90% argon/ 10% methane. Improved resolution (by a factor of about 6) can be achieved by using a Si(Li) detector mounted very close to the specimen surface in a snout arrangement devised by Cairns et al. (1970). One of the disadvantages associated with the use of the solid state detectors is that they are operated near liquid nitrogen temperatures and so the active face of the detector must be protected from water vapour condensation by mounting the detector behind a gas tight beryllium window. This will attenuate the X-ray signal, but with optimised electronics a substantial improvement in resolution is possible. A practical illustration of the use of 30–60 keV Ne^+ ions to identify sulphur as a surface containment on copper, nickel and stainless steel is given by Cairns (1973). As a result of the identification of sulphur in trace amounts on these surfaces, Cairns was able to conclude that the sulphur arises as a result of out-diffusion from the metallic specimens during high temperature (800°C) annealing. Sputtering reduced the amount of sulphur on the surface and exposed more surface metal atoms. The complicating effects of sulphur as a surface impurity which can be altered by thermal cycling and sputter treatments is described also in the Auger studies by Bishop et al. (1971) and Jenkins and Chung (1971) which are discussed in the following section.

8.7 ANALYSIS BY ELECTRON IRRADIATION AND SPUTTERING

Nuclear and X-ray methods capable of yielding physical information of atomic species have been accompanied by the development of surface analytical processes for determining the chemical association of surface atoms. Although electron beams have been employed to examine surfaces for over 20 years, it is only within the last decade that such experimental methods have reached the state of highly specific surface analytical techniques and the proliferation of publications and conferences on surface examination alone is evidence of the rapid acceptance of the derivatives of low energy electron diffraction (LEED) or Auger electron spectroscopy (AES). It is the purpose of this section to outline some of the main features of the range of instrumentation along with modern methods of analysis using ion beams for sputtering. In addition to the general references cited (Kane and Larrabee, 1974, Mayer and Tu, 1974) there are excellent review articles on the comparison of various techniques from which further details of specific methods can be obtained (e.g. Colligon, 1974; Coburn and Kay, 1974; Rivière, 1975).

There is a natural distinction between techniques which use sputtering to remove surface layers in order to examine the underlying material and those methods for which the sputter-ejected materials is itself analysed. The former category comprises such methods as AES and, more recently, XPS (X-ray Photoelectron Spectroscopy), whereas in SIMS (Secondary Ion Mass Spectroscopy) and SCANIR (Surface Compositional Analysis by Neutral and Ion Impact Radiation) the sputtered particles are monitored after they have been removed from the surface. We begin with developments from analytical methods based originally on LEED equipment (Weber and Peria, 1967) which fall under the general terminology of ESCA (Electron Spectroscopy for Chemical Analysis). The acronym ESCA is now applied to all surface analytical methods for which ejected secondary electrons are studied. Most experimental arrangements for ESCA employ Auger electrons. In Auger electron spectroscopy a sample is mounted in a high vacuum

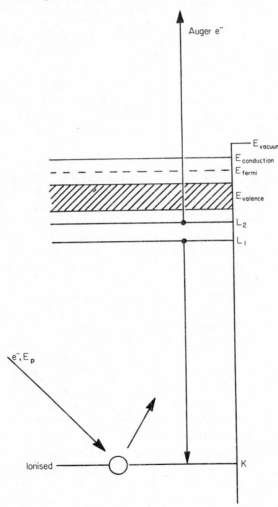

FIG. 8.27. Electron energy level diagram illustrating a transition giving rise to the emission of a KL_1K_2 Auger electron (after Stein et al., 1971).

(10^{-8} Torr) and irradiated with monoenergetic electrons of between 2 and 5 keV. In AES it is usually assumed that the excitation of inner shell electrons within the surface occurs due to electron beam irradiation, although Auger electrons can be observed also as a result of photon or ion bombardment. The Auger process is as follows. Within a short time (10^{-15} s) after excitation an inner shell hole is filled by an electron dropping down from an outer shell. The energy lost by this replacement electron is released either in the form of a photon, which constitutes the observed signal in electron induced X-ray analysis (XPS) or the energy is transferred to another electron (Auger electron) in an outer shell which is subsequently ejected from the atom (Figure 8.27). A small fraction (10^{-4}) of the Auger electrons escape from the solid and are detected by an electron energy analyser. If the initial vacancy-hole was in the K shell, the electron filling the hole came from the L_1 shell and if the Auger electron originated in the L_2 shell (Fig. 8.27) then the Auger electron is designated KL_1L_2. Auger electrons have an energy characteristic of the element from which they originated. The energy of a KL_1L_2 Auger electron is given by

$$E_{Ae} = E_K - E_{L_1} - E_{L_2} \qquad (8.24)$$

with appropriate corrections for relaxation effects and work function differences (Chung and Jenkins, 1970). The corrections are necessary in order to account for the doubly ionised state of the atom since there are vacancies in the L_1 and L_2 shells after irradiation. Tables of characteristic Auger electron energies are available and enable elements to be identified directly. During irradiation by a monoenergetic beam, secondary electrons are reflected, scattered and ejected from the surface at all energies from zero to E_p. The spectrum which results from such bombardment is characteristic not only of the surface, but of its chemical state also, Rivière (1973). The Auger peaks on the background spectrum can be observed more easily if the energy density curve $N(E)$ is differentiated $dN(E)/dE$ (Fig. 8.28). AES is a highly sensitive surface detection method; an equivalent surface concentration of 10^{-3} monolayers can be observed (Benninghoven, 1973). The sensitivity of AES to small amounts of adsorbed oxygen enabled Bishop *et al.* (1971) to establish through an extensive study on titanium that oxygen-free spectra can be obtained from this metal only above 780°C. At lower temperatures other contaminants such as sulphur, carbon and chlorine were detected.

By sputtering a sample with Ar^+ ions between successive AES probes it is possible to achieve a fair degree of depth resolution in association with the compositional data from the Auger spectra. Behrisch *et al.* (1973) achieved a depth resolution of 50 Å with AES combined with sputter removal of the surface atoms of thin film target specimens, and used the chemical sensitivity

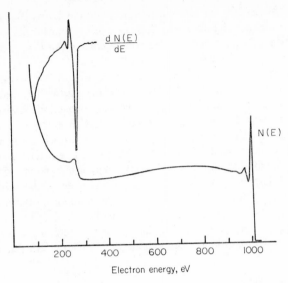

FIG. 8.28. Auger electron data from a carbon surface at a primary electron energy of 1 keV. Upper curve is a portion of the differential distribution $dN(E)/dE$ demonstrating the improved resolution of Auger features gained by differentiating with respect to energy (Rivière, 1973).

of AES to complement a Rutherford backscattering study of the films. Hartley and Coad (1975) have examined thick carbonaceous films on steel containing out-diffused sulphur and implanted nitrogen using XPS together with sputter removal in 20–40 Å intervals. Chemical information is obtained from AES because the binding energy of atomic inner shell electrons is influenced to a small extent by the chemical bonding of an atom. This shift in binding energy can be observed by both AES and XPS. However in AES the shifts of all three electronic levels involved in the Auger transition is reflected in the observed energy of emission, whereas in XPS the chemical shift of only the level involved in the transition is measured directly. Consequently XPS is much preferred as a probe for determining with precision the electronic energy levels in solids. AES does suffer from matrix effects and an example of this is the examination of insulating surfaces such as SiO_2 containing impurity chlorine. Flooding the surface with electrons at around 2 keV causes electron stimulated desorption of Cl (Chu et al., 1973). In addition changes in the surface topography result for materials such as the alkali halides which dissociate under electron bombardment (e.g. Elliott and Townsend, 1971).

8.7 ANALYSIS BY ELECTRON IRRADIATION AND SPUTTERING

In XPS a soft X-ray photon of known wavelength, usually Al or Mg K_α radiation, is directed at the specimen surface. The beam photoionises some of the atoms on the surface by the ejection of an inner shell electron. The energy distribution of the ejected photoelectrons is measured and since the electron has received a discrete quantum of energy (hv), the binding energy of the electron in the solid can be determined if the photoelectron escapes from the surface without suffering energy losses (Coburn and Kay, 1974). If the effective work function of the analyser (Fig. 8.29) is ϕ_a then the kinetic energy E_{kin} measured by the analyser is

$$E_{kin} = hv - E_{binding} - e\phi_a. \tag{8.25}$$

The identity of the element responsible for the photoelectron can be uniquely determined from the binding energy, and the intensity of the photoelectron current is a measure of the concentration of the element in the solid. XPS analysis on ion implanted steels has been used to identify the presence of

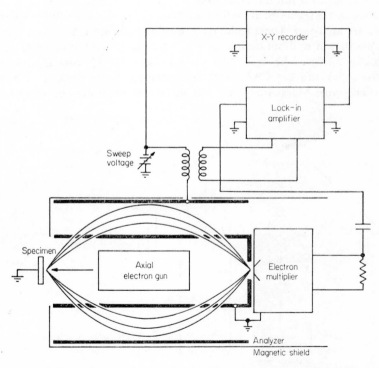

FIG. 8.29. Schematic layout of apparatus for AES using a cylindrical mirror analyser (Stein et al., 1971).

sulphide radicals on Mo and S implanted surfaces and friction-induced $Pb_2 O_3$ on lead implanted samples (Hartley et al., 1973, 1974). XPS is more favoured for chemical analysis since the information in this case is provided by the electron which was ejected from the inner shell by the incident radiation, while in AES the ejected electron provides no directly useful information because it does not absorb a known amount of energy from the incident electron flux (Coburn and Kay, 1974). Auger peaks are observed in XPS data and may be distinguished from the photoelectron peaks by using a different incident photon energy. The energy at which Auger peaks occur does not depend on the incident energy whereas the energy at which XPS peaks occur is directly proportional to the energy of the incident radiation. Published work in which AES or XPS data on ion implanted samples is included seems still to be sparse. As implantation is applied increasingly to metals for chemical reasons and corrosion studies it will probably become more usual to apply electron/photon analysis methods, in combination with sputtering removal, to probe a variety of unique systems in what is a rapidly expanding field for ion implantation.

The second category of analytical methods are those in which the sputtered species are analysed. These techniques are therefore more directly applicable to implantation and sputtering studies because of their inherently greater depth capabilities. The most well known are SIMS (Secondary Ion Mass Spectroscopy) and the CAMECA ion probe (or its equivalent—IMMA—Ion Microprobe Mass Analyser). A relatively recent system of analysis with

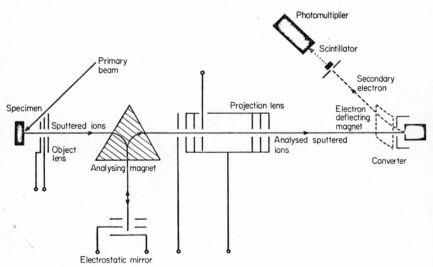

FIG. 8.30. Schematic diagram of the CAMECA ion microanalyser (Gittins et al., 1972).

the acronym SCANIIR (Surface Compositional Analysis by Neutral and Ion Impact Radiation) has been developed and this is discussed below since its versatility and application to insulating materials may provide important new fundamental information on the energy level states of complex samples.

The emission of both positive and negative ions from bombarded surfaces has been studied for several decades, but it is only in the last 12 years since the development of sophisticated methods such as the CAMECA instrument (Castaing and Slodzian, 1962) that ionic emission has achieved the status of an analytical tool. A schematic diagram of CAMECA is given in Fig. 8.30. A primary ion beam (usually oxygen or argon) of energy 6–20 keV is focussed onto a specimen in an evacuated chamber at $5 \cdot 10^{-8}$ Torr. With a probing beam current of about 15 μA it is possible to achieve a mass analysis spectrum in a few minutes, to a sensitivity of better than 1 ppm in most cases. Secondary ions from the sample are extracted at energies of typically 5 keV and are mass analysed in a stigmatic magnetic prism. They are then reflected and energy analysed by an electrostatic mirror lens which serves to finely select ions of interest to fractions of an eV; the ions are subsequently mass analysed and detected either by visual observation as a spectrum accumulates or by using a scintillation counter and photomultiplier detector system so that a display of the relevant species can be monitored on a screen (Castaing, 1972). If a depth distribution, e.g. of implanted boron in silicon is required then the CAMECA is obviously applicable to such studies since the surface is eroded by sputtering during observation. Thus Gittins *et al.* (1972) established the implantation profile of boron in silicon and demonstrated that the instrument is suitable for this purpose if the sample sputtering rate is calibrated against a similar specimen of known dopant concentration. Tsai *et al.* (1973) also used the CAMECA instrument on arsenic implanted silicon, and their results agreed well with theoretical predictions for the range of As in Si. One possible source of error with the CAMECA system, and techniques which employ sputtering to eject surface species, is the appearance of "tails" in the distribution due to "knock-on" effects from the primary beam. Tsai *et al.* conclude that although such tails are not significant in their experiments they could be due to preferential erosion of the sputter-induced crater walls on the samples. Such spurious results may be minimised by selecting ions from a central region of the specimen surface damaged by the beam (Gittins *et al.*, 1972). The sensitivity of CAMECA is very good, up to better than 1 ppm, which means that implantation doses of between 10^{14} and 10^{15} ions cm^{-2} may be detected, provided their sputtering rates are sufficiently high for the particular primary beam employed. However one major disadvantage associated with this machine is its cost ($\sim$ £100,000) and this is one reason why interest has been focussed onto the development of more compact systems such as SIMS. The SIMS arrangement also examines ion-ejected species, but in this case much less finely focussed beams are employed (i.e. several mm width as opposed to ~ 100 μm) at lower currents (~ 100 nA) so that effectively the sputtering rate is zero. In SIMS a mass-analysed primary ion beam of (usually) gaseous ions at 0·1–1 keV bombards the sample at 70° to the normal to the surface plane. Residual gas pressures are about 10^{-11} Torr, rising to 10^{-8} Torr of argon when the source operates. Secondary ions are extracted normal to the sample surface and are analysed by a quadrupole mass filter which focusses ions of a particular mass through an aperture where they are deflected into a channeltron multiplier and detected by pulse counting techniques (Fig. 8.31). The quadrupole system

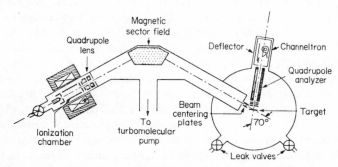

FIG. 8.31. Arrangement for Secondary Ion Mass Spectroscopy (Benninghoven and Loebach, 1971).

has the advantage of being simple and easily bakeable, in addition to having good mass resolution. Since in SIMS the surface itself is removed and mass analysed the technique provides both compositional and chemical information. All elements present on the surface at concentrations above the limits of sensitivity (1 part in 10^9) will be observed, including hydrogen and molecules containing hydrogen that cannot be detected by other surface analytical methods. However there is the major disadvantage, alluded to earlier, associated with the analysis of sputter ejected systems in that the sputtering yield can vary over three orders of magnitude for elements in the pure state (Fig. 8.32), and are profoundly affected by the chemical environment of a particular species. Thus the emission of an element from its oxide can exceed that from the pure state, an effect attributed to effective changes in the work function depending on the electronegativity of the bombarding particles (Anderson and Hinthorne, 1973). An alternative explanation is proposed by Blaise and Slodzian (1968) who discuss the self-ionisation of highly excited neutral ions ejected from the surface. The secondary ionisation process is highly complex and the reader is referred to the review articles of Coburn and Kay (1974), Colligon (1974) and a recent discussion of

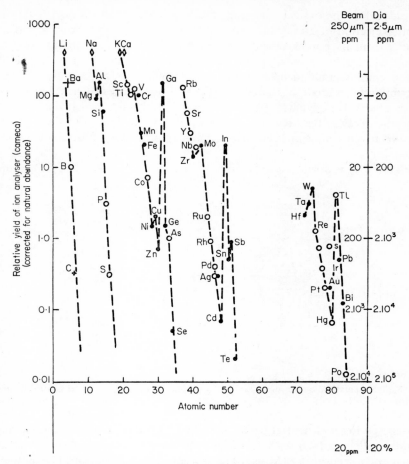

FIG. 8.32. Relative yield of the CAMECA ion analyser for various sputtered ions, corrected for natural abundance of isotope studied (Gittins et al., 1972).

Werner (1975) for further references and details of theoretical models. Hofker et al. (1973) applied a SIMS analysis to boron implanted silicon and attributed observed concentration tails to channelling. SIMS has been applied to alkali halides, principally to determine the effects of high secondary ionisation efficiencies on surface emission rates (Slodzian and Hennequin, 1966).

SCANIIR (White and Tolk, 1971) combines the sensitivity of SIMS with the energy level selectivity of AES and has been applied with considerable success to both metallic and insulating substrates. The method can be considered as the optical analogue of SIMS (Coburn and Kay, 1974). The sputter ejection of atoms and molecules in an excited state from a surface results from its bombardment with nitrogen or inert gas ions at energies < 10 keV. Instead of being passed through a mass analyser, the SCANIIR system employs a monochromator and photomultiplier to record the spectral distribution of radiation produced in the collision process. Optical emission lines, observable as narrow peaks on a relatively flat background, result from the decay of the excited emission states, and the lines provide the basis for a relatively simple and sensitive technique for the identification of surface species.

Figure 8.33 is a schematic diagram of the apparatus used to produce the

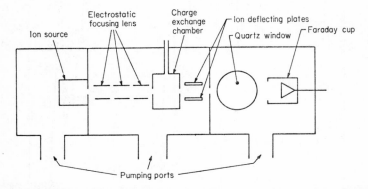

FIG. 8.33. Schematic diagram of apparatus to produce low energy ion and neutral beams used in SCANIIR (White *et al.*, 1974).

low-energy ion and neutral beams used in SCANIIR. The charge exchange chamber allows a focussed beam of ions to become neutralised by means of resonant charge transfer (Massey and Burhop, 1952). The neutral beam then impinges on the target with essentially the same kinetic energy as the original ion beam. In certain situations it is not necessary to neutralise the beam and the ion beam is focussed directly through to the sample surface. Spectral analysis of the emitted radiation is achieved by scanning the monochromator through the visible wavelength range, 2000–8000 Å. This spectra can then be related to the elemental species originally present on the surface or, in certain cases, ascribed to electronic transitions arising within the bulk material (White *et al.*, 1974). The intensity of a single line as a function of beam energy or bombarding time using either a monochromator or interference filter may also be monitored to assess surface composition or depth profiling, with the usual constraints of sputtering efficiency. The various collision-induced optical radiation peaks which have been identified result from the excitation of surface sputtered and backscattered particles. One of the advantages of SCANIIR is that insulating materials may be examined (without the problems associated with charge build-up on the specimen surface) by the use of neutral probing beams. White et al. (1974) discuss the spectral distribution of visible radiation from neutral nitrogen (N_2^0) impact on Al_2O_3, LiF and SiO_2 and demonstrate that a substantial proportion of the sputtered material is ejected as individual excited-state atoms. Unfortunately elements such as O and F which radiate primarily in the ultraviolet region of the spectrum cannot be detected by SCANIIR without experimental modifications. If a metal or semiconducting target is examined, neutrals and ions of the same species are found to produce photons with equal efficiency because at low energies an ion is neutralised several Ångströms in front of the surface by non-radiative de-excitation (i.e. Auger-type processes), Hagstrum (1961). Auger emission induced during SCANIIR probing has itself been applied to the analysis of materials such as SiO_2 although whether the de-excitation is one-electron (similar to XPS processes) or two-electron (Auger) depends on the energy levels of the solid under investigation and the distance of closest approach of the probing beam at which an encounter-induced electron emission occurs, van der Weg and Bierman (1969). An outline of the theoretical models relevant to de-excitation processes under these conditions is given by White et al. (1974) and by Colligon (1974).

SCANIIR has been applied to a variety of complex solids including geological samples, metallic systems, alkali halides and liquid residues such as blood plasma (White et al., 1972). The detection sensitivities are normally in the order of 1–10 ppm although substantial improvements are considered possible with the use of narrow-band interference filters instead of monochromators to isolate prominent optical lines emitted by sputtered impurity atoms. Profiling measurements in $SiO_2 - Al_2O_3$ sandwich targets have shown that in principle the rate of change of composition of a surface with depth can be determined, but the usual uncertainties in sputtering rates make quantitative analyses difficult. Optical radiation is observed also to arise from the excited states of backscattered incident beam particles which have escaped the surface after large angle collisions have occurred within the bulk material. Studies of backscattered radiation provide more complete information about final states than do conventional charge-state measurements (White et al., 1974). Radiation arising from backscattered particles is shown in Fig. 8.34 which shows the measured line profile of neutral He–5876 Å emission produced by the impact of He^+ (90 keV) on copper. This emission line is substantially Doppler-shifted and broadened, which indicates the wide range of energies and direction of backscattered projectiles as well as effects due to non-radiative de-excitation. When related to energy level structure such data can be used to determine the population of excited states for backscattered particles and it should in principle be possible to relate the spectral shifts to the relative role of ejection from the bulk material or surface.

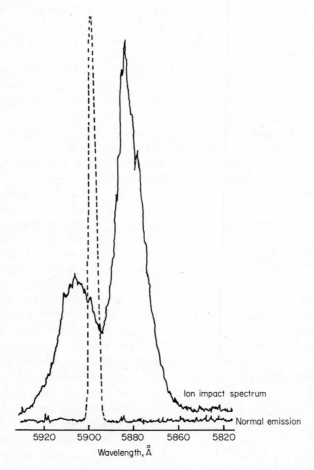

FIG. 8.34. Measured line profiles of neutral He (HeI-5876 Å) produce by the impact of He$^+$ at 90 keV on copper. The position of the unshifted line and the instrumental resolution is indicated by the dotted line (White et al., 1974).

In alkali halides collisions within the bulk crystal can result in ejection of both photons and atoms (see, for example, Townsend and Kelly, 1973). An example is given in Fig. 8.35 which shows the spectral distribution of radiation produced from the impact of He^0 (5 keV) on single crystal CaF_2. The collisions of neutral helium with the substrate produce a continuum of radiation extending over about 1800 Å with the broad peak centred around 2800 Å. The narrow prominent peaks result from the excited states of Ca and Ca^+. Similar continuum radiation is produced in low energy electron bombardment on the same material (White *et al.* (1974) and Hayes *et al.* (1969)) and these latter authors ascribe the continuum to radiative recombination of electrons with the self-trapped V_K centre (an F_2^- molecule). White *et al.* attribute the continuum to radiative recombination of the self-trapped hole and mobile electrons created in parallel with holes in the elastic interaction of the projectile (helium) with electrons in the crystal. Radiative continua are produced as a result of collisions of ions with metal targets such as Ni, Ta and Mo (Tolk *et al.*, 1973; McCracken and Erents, 1970). The reason for the difference between ion–metal and ion–insulator continuum spectra remains to be established, but the development of SCANIIR ensures its application to studies of energy levels from surface and bulk targets and possibly therefore to the detailed experimental investigation of quasi-binary energy loss mechanisms in solids.

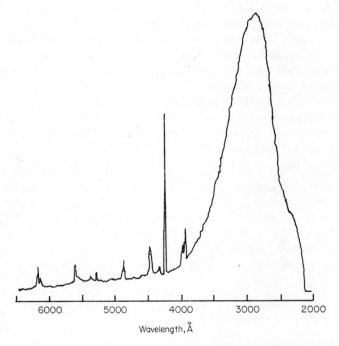

FIG. 8.35. SCANIIR spectrum of radiation produced in the impact of He^0 (5 keV) on CaF_2 (White *et al.*, 1974).

Other techniques which employ ion and neutral beams are Ion Scattering Spectroscopy (ISS) and Ion Neutralisation Spectroscopy (INS). INS is closely related to SCANIIR since if an ionised atom approaches a solid surface sufficiently slowly, a metastable excited ion-atom system is formed within a certain critical distance from the surface (Hagstrum, 1966 and 1970) and the theory is as before. In practise INS probes the first 1–4 monolayers in common with AES since de-excitation occurs by the emission of an electron from the valence band somewhere in this region. The important condition which must be fulfilled if this type of auto-ionisation is to take place is that there should not be any atomic levels whose energies lie in the energy range of the filled band of the solid (Rivière, 1975). For this reason INS can be considered to be a filled band probe, as can AES, and may be used to examine energy levels at the "surface" of solids. INS is most conveniently carried out by bombarding a solid with He^+ ions in the ground state at energies between 0 and 20 eV. Suitable target materials are solids whose filled valence band lies entirely between 4·5 and 22·5 eV which fortunately includes most of the metal oxides. INS is not a compositional technique since the resultant spectra contains no Z_2 dependent information as in the case with XPS and AES. INS generates spectroscopic information on the surface electronic states and therefore produces chemical information about the methods by which the filled bands of solids are altered at the surface as a consequence of chemical reaction (Rivière, 1975). Thus Becker and Hagstrum (1972) studied the adsorption states of a clean nickel surface to which sub-monolayer amounts of H_2S were admitted. They demonstrated that the surface density of states for sulphur and adsorbed oxygen depends on the symmetry of the surface molecular orbitals as well as on the nature of the adsorbed atom, since the d-band population for each species varied according to the crystallographic plane of the substrate nickel.

Ion Scattering Spectroscopy is a truly surface specific analysis method capable of detecting about $\sim 0.1\%$ of a monolayer under favourable conditions. The experimental process consists of bombarding the sample with a beam of ions, usually He^+, with an energy in the range 0·5 to 2 keV and recording the energy distribution of the backscattered projectile ion at a fixed scattering angle, usually 90° (Smith, 1967). The particles are collected by an energy analyser and the energy distribution has sharp peaks at energies E_2 where

$$\frac{E_2}{E_1} = \frac{M_2 - M_1}{M_2 + M_1} \tag{8.26}$$

(with the usual notation) for 90° scattering, in an exactly analgous process to Rutherford backscattering. The intensity of the peaks at Z_2 is a measure of

8.7 ANALYSIS BY ELECTRON IRRADIATION AND SPUTTERING

the surface coverage of M_2. Ideally, each collision is an elastic two-body process on the outermost monolayer. Penetration of the beam produces broadening of the peaks and additional considerations have to be introduced into the scattering theory to account for collisions in which the deflected particles interact with more than one surface atom. Since the interaction in ISS is with the nuclei of the surface, such binary collision events will provide energy spectra characteristic of the masses of two scattering centres regardless of the conductivity of the specimen. A good review of ISS and an extensive bibliography relevant to the subject is given by Honig and Harrington (1973), from which Fig. 8.36 is taken. This shows the spectrum of 1·5keV $^{20}Ne^+$

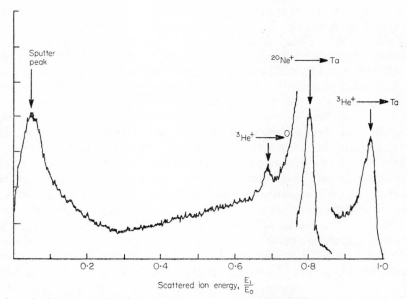

FIG. 8.36. ISS spectrum of 1·5 keV $^{20}Ne^+$ ions scattered from Ta_2O_5, (Honig and Harrington, 1973).

ions scattered from Ta_2O_5, and illustrates the sensitivity of the method which depends in a complex way on matrix effects and geometry to give quoted sensitivities between 10^{-4} and 10^{-1} monolayers. Figure 8.36 also shows the presence of a sputter peak which severely limits the detection of low mass components. Nevertheless, for outermost monolayer surface studies on materials such as Ta_2O_5 and other semiconductors and insulators, ISS is a particularly convenient method of analysis since oxygen-enhanced elemental sensitivities frequently encountered in SIMS are absent. In addition, the scattered beam intensity is a function of lattice geometry and can therefore be used to identify a given crystal face. Thus ISS is the elastic (nuclear) analogue of INS and may provide important complementary data on physical and chemical properties at the monolayer level for a variety of solid surfaces of interest to such fields as catalysis, corrosion and adsorption studies on metals and insulators.

8.8 CONCLUSIONS

The techniques which have been described in this chapter cover a range of sensitivities from a few percent to < 0·1 ppm. Their application extends across a variety of materials and it has been intended in the previous sections to give examples of the type and quality of information obtainable from each process. Clearly the optimum analytical method for a particular sample will depend on the suitability of the specimen to destructive testing, for example, or on its physical and chemical nature as well as on the availability of equipment. Access to a particle accelerator places Rutherford backscattering and nuclear reaction analysis within reach of the analyst so that rapid compositional or isotopic data can be generated over a depth of a few microns. However, the limited depth resolution of such methods will in general preclude their application to surface sensitive substrates and adsorption studies at the monolayer level of interest, where ISS and INS become applicable. Channelling methods in metals and insulators require specialist experience to gain maximum advantage from the unique quality of data inherent to such experiments and at present channelling is confined to fundamental materials research rather than batch analysis. Highest elemental sensitivities are achieved with SIMS and SCANIIR and since secondary ion emission systems are becoming more widespread greater use will be made of such capabilities. Provision has to be made for non-uniform sputtering (cratering) and matrix effects, such as the enhanced elemental ejection due to local energy level perturbations brought about by the primary beam. XPS and Auger Electron Spectroscopy are only just beginning to be used in conjunction with ion implanted and sputtered surfaces and it would seem that their application to samples modified in these ways for material property changes is particularly apt provided the underlying electronic transitions may be accounted for. But these techniques also have matrix effects, as indicated in the previous section, which are most pronounced in alkali halides and insulators. Ion induced X-rays are suited to surfaces in which specific elements are found whose X-ray generation can be enhanced by the appropriate choice of bombarding energy, but although the basic physical processes of heavy ion induced X-rays are understood there still remain areas of incomplete theory and so the general application of the techniques is limited to simple systems. Nevertheless there is an essential role for experimental methods which provide data for comparison with theoretical models. The most convenient and rapid multielement method for comparative work is probably proton induced X-ray analysis (PIXA) but again the high energy accelerators required for this work limits its availability. The physical information generated by SCANIIR and its applicability to a variety of materials makes this system attractive as well as providing considerable data for theoretical treatment such as the electron ejection and excited-state sputtered particle data discussed in the previous section. But SCANIIR as an analytical tool is not as generally available as some of the more well-known techniques such as SIMS, AES and XPS. In conclusion, the benefits of a particular technique are often not appreciated until a comparison is made with other methods. As a result it is frequently instructive to complement analysis techniques on the same sample.

REFERENCES

Abel, F., Bruneaux, M., Cohen, C., Chaumont, J. and Thom, é. J. (1973a). *Solid State Comm.* **13**, 113.
Abel, F., Amsel, G., d'Artemare, E., Bruneaux, M., Cohen, C., Maurel, B., Ortega, C., Riga, S., Siejka, J., Croset, M. and Dieumegard, D. (1973b). *J. Radioanal Chem.* **16**, 567.
Abel, F., Bruneaux, M., Cohen, C., Bernas, H., Chaumont, J. and Thomé, L. (1973c). *Sol. State Comm.* **13**, 113.
Agius, B., Siejka, J., Amsel, G. and Gibson, W. M. (1975). To be published.
Ajzenberg-Selove, F. and Lauritsen, T. (1968). *Nucl. Phys.* **A114**, 1.; 1974, *ibid*, **A227**, 1.
Akasaka, Y. and Horie, K. (1973). In "Ion Implantation in Semiconductors and Other Materials", (Ed. Crowder), Plenum, p. 147.
Alexander, R. B., Dearnaley, G., Morgan, D. V., Poate, J. M. and Van Vliet, D. (1970). In "Proc. Europ. Conf. on Ion Impltn.", Peregrinus, p. 181.
Alexander, R. B., Callaghan, P. T. and Poate, J. M. (1973). In "Ion Implantation in Semiconds. and Other Materials" (Ed. Crowder), Plenum, p. 477,.
Alexander, R. B., Ansaldo, E. J., Deutch, B. I., Gellert, J. and Feldman, L. C. (1974). In "Proc. Internat. Conf. on Applicns. of Ion Beams to Metals". (Ed. Picraux *et al.*), Plenum, p. 365.
Allen, C. R., Pritchard, A. M. and Hartley, N. E. W. (1975). Unpublished.
Amsel, G., Nadai, J. P., D'Artemare, E., David, D., Girard, E. and Moulin, J. (1971). *Nucl. Instr. Methods,* **92**, 481.
Anderson, C. A. and Hinthorne, J. R. (1973). *Anal. Chem.* **45**, 1421.
Appleton, B. R., Erginsoy, C. and Gibson, W. M. (1967). *Phys. Rev.* **161**, 330.
Baglin, J. E. E. (1975). Int. Conf. on Ion Beam Surface Layer Analysis, Karlsruhe. To be published.
Ball, D. J., Buck, T. M., MacNair, D. and Wheatley, G. H. (1972). *Surf. Sci.* **30**, 69.
Barnes, D. G., Calvert, J. M., Kay, K. A. and Lees, D. G. (1973). *Phil. Mag.* **28**, 1303.
Barragan, A. (1974). Unpublished.

Bashkin, S., Carlson, R. R. and Douglas, R. A. (1959). *Phys. Rev.* **114,** 1552.
Becker, G. E. and Hagstrum, H. D. (1972). *Surf. Sci.* **30,** 505.
Behrisch, R., Scherzer, B. M. U. and Staib, P. (1973). *Thin Solid Films,* **19,** 57.
Benninghoven, A. (1973). *Appl. Phys.* **1,** 3.
Benninghoven, A. and Loebach, E. (1971). *Rev. Sci. Instr.* **42,** 49.
Bird, J. E., Campbell, B. L. and Price, P. B. (1974). *Atomic Energy Rev.* **12,** 275.
Bishop, H. E., Rivière, J. C. and Coad, J. P. (1971). *Surf. Sci.* **24,** 1.
Bisson, C. L. and Wilson, W. D. (1974). *In* "Proc. Internat. Conf. on Applcns. of Ion Beams to Metals" (Ed. Picraux *et al.*), Plenum, p. 423.
Blaise, G. and Slodzian, G. (1968). *C.R. Acad. Sci. Paris Ser. B* **266,** 1525.
Blood, P., Dearnaley, G. and Wilkins, M. A. (1974). *J. Appl. Phys.* **45,** 5123.
Bøgh, E. (1968). *Can. J. Phys.* **46,** 653.
Buck, T. M. and Wheatley, G. H. (1972). *Surf. Sci.* **33,** 35.
Cairns, J. A., Desborough, C. L. and Holloway, D. F. (1970). *Nucl. Instr. and Methods,* **88,** 239.
Cairns, J. A. (1973). *Surf. Sci.* **34,** 638.
Cairns, J. A., Marwick, A. D. and Mitchell, I. V. (1973). *Thin Solid Films,* **19,** 91.
Calvert, J. M., Derry, D. J. and Lees, D. G. (1974). *J. Phys. D.* **7,** 940.
Cameron, J. R. (1953). *Phys. Rev.* **90,** 839.
Campisano, S. U., Foti, G., Grasso, F., Mayer, J. W. and Rimini, E. (1974). *In* "Proc. Internat. Conf. on Applcns. of Ion Beams to Metals". (Ed. Picraux *et al.*), Plenum, p. 159.
Carstanjen, H. D. and Sizmann, R. (1972). *Phys. Lett.* **40A,** 93.
Case, B., Allen, C. R., Bradford, P., Dearnaley, G., Turner, J. F. and Woolsey, I. S. (1975). Presented at Conf. on Ion Implantation and Ion Beam Analysis Techniques, UMIST.
Castaing, R. and Slodzian, G. (1962). *J. de Microscopie,* **1,** 395.
Castaing, R. (1972). *In* "Proc. 6th Internat. Conf. on X-ray Optics and Micro Anal." (Ed. Shinoda *et al.*), p. 399.
Chemin, J. F., Mitchell, I. V. and Saris, F. W. (1973). *In* "Ion Implantation in Semiconds. and Other Materials. (Ed. Crowder), Plenum, p. 295.
Chu, W. K., Mayer, J. W., Nicolet, M-A, Buck, T. M., Amsel, G. and Eisen, F. (1973). *Thin Solid Films,* **17,** 1.
Chung, M. F. and Jenkins, L. H. (1970), *Surf. Sci.* **22,** 479.
Clark, G. J., Morgan, D. V. and Poate, J. M. (1970). *In* "Atomic Collis, Phenom. in Solids" (Eds Palmer, Thompson and Townsend), North-Holland, p. 388.
Coburn, J. W. and Kay, E. (1974). *Crit. Rev. Solid State Sci.* **4,** 561.
Colligon, J. S. (1974). *Vacuum,* **24,** 373.
Cookson, J. A. (1975). Unpublished
Cookson, J. A. and Pilling, F. D. (1970). UKAEA Rept. AERE R6300.
Cookson, J. A. and Pilling, F. D. (1973). *Thin Solid Films,* **19,** 381.
Crowder, B. L. (Ed.), (1973). "Ion Implantation in Semiconds. and Other Materials", Plenum, New York.
Darken L. S. and Gurry, R. W. (1953). "Physical Chemistry of Metals", McGraw-Hill.
Datz, S., Appleton, B. R. and Moak, C. D. (Eds) (1975). "Atomic Collisions in Solids", 5th Internat. Conf., Gatlinburg, U.S.A., Plenum Press, p. 142.
Davies, J. A. (1973), Chap. 13 in "Channelling" (Ed. Morgan), John Wiley.
Davies, J. A., Denhartog, J., Eriksson, L. and Mayer, J. W. (1967). *Can. J. Phys.* **45,** 4053.

Dearnaley, G., Goode, P. D., Miller, W. S. and Turner, J. F. (1972). *In* "Ion Implantation in Semiconds. and Other Materials." (Ed. Crowder), Plenum, p. 403.
Dearnaley, G. (1974). *In* "Applcns. of Ion Beams to Metals," (Ed. Picraux *et al.*), Plenum, p. 63.
Dearnaley, G., Garnsey, R., Hartley, N. E. W., Turner, J. F. and Woolsey, I. S. (1975). *J. Vac. Sci. Technol.*, **12,** 449.
Dearnaley, G. and Squires, G. (1975). Unpublished.
DeBonte, W. J., Poate, J. M., Melliar-Smith, C. M. and Levesque, R. A. (1974). *In* "Proc. Internat. Conf. on Applcns. of Ion Beams to Metals" (Ed. Picraux *et al.*), Plenum, p. 147.
Della Mea G., Dirgo, A. V., Lo Russo, S., Mazzoldi, P., Yamaguchi, S., Bentini, G., Desalvo, A. and Rosa, R. (1975). *In* "Internat. Conf. on Atomic Collsn. in Solids," (Ed. Datz *et al.*), Plenum, p. 791.
Der, R. C., Kavanagh, T. M., Khan, J. M., Curry, B. P. and Fortner, R. J. (1968). *Phys. Rev. Letters,* **21,** 1731.
Duggan, J. L., Beck, W. L., Albrecht, L., Munz, L. and Spaulding, J. D. (1972). *Advances in X-ray Analysis,* **15,** 407.
Elliott, D. J. and Townsend, P. D. (1971). *Phil. Mag.* **23,** 249.
Erents, S. K. and McCracken, G. M. (1973). *Rad. Effects,* **18,** 191.
Everhart, E., Stone, G. and Carbone, R. J. (1955). *Phys. Rev.* **99,** 1285.
Fano, U., and Lichten, W. (1965). *Phys. Rev. Letters,* **14,** 627.
Feldman, L. C., Kaufmann, E. N., Poate, J. M. and Augustyniak W. M. (1973). *In* "Ion Implantation in Semiconds. and Other Materials." (Ed. Crowder), Plenum p. 491.
Feldman, L. C., Poate, J. M., Ermanis F. and Schwartz, B. (1973). *Thin Solid Films,* **19,** 81.
Feng, J. S. Y., Chu, W. K. and Nicolet, M. A. (1973). *In* "Ion Implantation in Semiconds. and other Materials" (Ed. Crowder), p. 227.
Fortner, R. J., Currey, B. P., Der R. C., Kavanagh T. M. and Khan J. M. (1969). *Phys. Rev.* **185,** 164.
Folkmann, F., Gaarde, C., Huus, T. and Kemp, K. (1974). *Nucl. Instrs. Methods,* **116,** 487.
Garcia, J. D. (1970). *Phys. Rev. A.* **1,** 280; *ibid* 1402.
Garcia, J. D., Fortner, R. J. and Kavanagh, T. M. (1973). *Rev. Mod. Phys.* **45,** 111.
Gemmell, D. S. (1974). *Rev. Mod. Phys.* **46,** 129.
Gilfrich, J. V. (1974). Chap. 12 *In* "Characterization of Solid Surfaces" (Eds Kane and Larrabee), Plenum.
Gittins R. P., Morgan, D. V. and Dearnaley, G. (1972). *J. Phys. D.* **5,** 1654.
Gordon, B. M. and Kraner, H. W. (1971). BNL report, 16182.
Grasso, F. (1973). Chap. 7 *In* "Channelling" (Ed. Morgan), John Wiley, 181.
Gray, T. J., Lear, R., Dexter, R. J., Schwettmann, F. N. and Wiemer, K. C. (1973). *Thin Solid Films,* **19,** 103.
Gyulai, J., Meyer, O., Mayer, J. W. and Rodriguez, V. (1970). *Appl. Phys. Lett.* **16,** 232.
Hagstrum, H. D. (1961). *Phys. Rev.* **123,** 758.
Hagstrum, H. D. (1966). *Phys. Rev.* **150,** 495.
Hagstrum, H. D. (1970). *J. Vac. Sci. Technol.* **7,** 62.
Hartley, N. E. W., Dearnaley, G. and Turner, J. F. (1973). *In* "Ion Implantation in Semiconds. and Other Materials." (Ed. Crowder), Plenum, p. 423.

Hartley, N. E. W., Dearnaley, G., Turner, J. F. and Saunders, J. (1974). *In* "Applcns. of Ion Beams to Metals," (Ed. Picraux *et al.*), Plenum, p. 123.
Hartley, N. E. W. and Coad, J. P. (1975). Unpublished.
Hayes, W., Kirk, D. L. and Summers G. P. (1969). *Solid State Commun.* **7**, 1061.
Hill, R. W. (1953). *Phys. Rev.* **90**, 845.
Hink, W. (1972). *In* "Proc. 6th Internat. Conf on X-ray Optics and Microanal." (Ed. Shinoda *et al.*), p. 523.
Hofker, W. K., Werner, H. W., Oosthoek, D. P. and De Grefte H. A. M. (1973). *In* "Ion Implantation in Semiconds. and Other Materials." (Ed. Crowder), Plenum, p. 133.
Honig, R. E. and Harrington, W. L. (1973). *Thin Solid Films*, **19**, 43.
Hume-Rothery, W., Smallman, R. E., and Haworth C. W. (1969). "The Structure of Metals and Alloys", Inst of Metals, London.
Huus, T. and Day, R. B. (1953). *Phys. Rev.* **91**, 599.
Jenkins, L. H. and Chung, M. F. (1971). *Surf Sci.* **24**, 125.
Johansson, T. B., Akselsson, R. and Johansson, S.A.E. (1970). *Nucl. Instr. and Methods*, **84**, 141.
Kahn, J. M., Potter, D. L. and Worley, R. D. (1965). *Phys. Rev.* **139**, A1735; (1966). *J. Appl. Phys.* **37**, 564.
Kane, P. F. and Larrabee, G. R. (Eds) (1974). "Characterization of Solid Surfaces", Plenum Press.
Kessel, Q. C. and Fastrup, B. (1973). *In* "Case Studies of Modern Physics," Vol. 3, (Eds. McDowell and McDaniel), North-Holland, p. 137.
Knox, K. C. (1970). *Nucl. Instr. Methods*, **81**, 202.
Laubenstein, R. A., Laubenstein, M. J. W., Koester, L. J. and Mobley, R. C. (1951). *Phys. Rev.* **84**, 12.
Ligeon, E. and Guivarc'h, A. (1974). *Rad. Effects*, **22**, 101.
Marion, J. B. and Young, F. C. (1968). "Nuclear Reaction Analysis", North-Holland.
Massey, H. S. W. and Burhop, E. H. S. (1952). "Electronic and Ionic Impact Phenomena", Clarendon Press.
Matzke, Hj. (1966). *Phys. Stat. Sol.* **18**, 285.
Matzke, Hj, Davies, J. A. and Johansson, N. G. E. (1971). *Can. J. Phys.* **49**, 2215.
Mayer, J. W., Eriksson, L. and Davies, J. A. (1970). "Ion Implantation in Semiconductors", Academic Press.
Mayer, J. W. and Turos, A. (1973). *Thin Solid Films*, **19**, 1.
Mayer, J. W. and Tu, K. N. (1974). *J. Vac. Sci. Technol.* **11**, 86.
Mayer, J. W. and Ziegler, J. F. (Eds) (1974). "Ion Beam Surface Layer Analysis", Elsevier.
Meyer, O., Linker, G. and Kraeft, B. (1973). *Thin Solid Films*, **19**, 217.
McCracken, G. M. and Erents, S. K. (1970). *Phys. Letters*, **31A**, 429.
Möller, E. and Starfelt, N. (1967). *Nucl. Instr. Methods*, **50**, 225.
Morgan, D. V. (Ed.) (1973). "Channelling Theory, Observation and Applications", Wiley.
Morgan, D. V. (1974). *J. Phys. D.*, **7**, 653.
Musket, R. G. and Bauer W. (1973). *Thin Solid Films*, **19**, 69.
Neild, D. J., Wise, P. J. and Barnes, D. G. (1972). *J. Phys. D.* **5**, 2292.
Nicolet, M-A., Mayer, J. W. and Mitchell, I. V. (1972). *Science*, **177**, 841.
North, J. E. and Gibson, W. M. (1970). *Appl. Phys. Letters*, **16**, 126.
Northcliffe, L. C. and Schilling, R. F. (1970). *Nucl. Data Sect.* **A7**, 233.

Olivier, C. and Pierce, T. B. (1974). *Radiochem. Radional. Letters*, **17**, 335.
Palmer, D. W., Thompson, M. W. and Townsend, P. D. (Eds) (1970). "Atomic Collision Phenomena in Solids", North-Holland.
Picraux, S. T., Brown, W. L. and Gibson, W. M. (1972). *Phys. Rev.* **B6**, 1382.
Picraux, S. T. and Vook, F. L. (1974). *In* "Applicns. of Ion Beams to Metals" (Ed. Picraux *et al.*), Plenum, p. 407.
Picraux, S. T., Eernisse, E. P. and Vook, F. L. (Eds) (1974). "Application of Ion Beams to Metals", Plenum Press.
Pierce, T. B., McMillan, J. W., Peck, P. F. and Jones, I. G. (1974). *Nucl. Instr. and Methods*, **118**, 115.
Pierce, T. B. (1974). Chapter 17, *In* "Characterization of Solid Surfaces," (Eds Kane and Larrabee), Plenum, p. 419.
Poate, J. M., DeBonte, W. J., Augustyniak, W. M. and Borders, J. A. (1974). *Appl. Phys. Letters*, **25**, 698.
Rickards, J. and Dearnaley, G. (1974). *In* "Applcns. of Ion Beams to Metals," (Ed. Picraux *et al*), Plenum, p. 101.
Rivière, J. C. (1973). *Contemp. Phys.* **14**, 513.
Rivière, J. C. (1975). To be published.
Rubin, S., Passell, T. O. and Bailey, L. E. (1957). *Anal Chem.* **29**, 736.
Rudd, M. E. (1972). *In* "The Physics of Electronic and Atomic Collisions", (Eds Govers and De Heer), North-Holland, p. 107.
Saris, F. W. (1972). *In* "The Physics of Electronic and Atomic Collisions", (Eds Govers and De Heer), North-Holland, p. 181.
Sattler, A. R. and Dearnaley, G. (1967). *Phys. Rev.* **161**, 244.
Sattler, A. R. and Vook, F. L. (1968). *Phys. Rev.* **175**, 526.
Schmid, K. and Ryssel, H. (1974). *Nucl. Instr. and Methods*, **119**, 287.
Slodzian, G. and Hennequin, J. F. (1966). *C.R. Acad. Sci. Paris Ser. B.* **263**, 1246.
Smith, D. P. (1967). *J. Appl. Phys.* **38**, 340.
Sood, D. K. (1975). *J. Vac. Sci. Technol.*, **12**, 463.
Stein, D. F., Weber, R. E. and Palmberg, P. W. (1971). *J. Metals*, **23**, 39.
Tolk, N. H., White, C. W. and Sigmund, P. (1973). *Bull. Am. Phys. Soc.* **18**, 686.
Townsend, P. D. and Kelly, J. C. (1973). "Colour Centres and Imperfections in Insulators and Semiconductors", Sussex Univ. Press.
Tsai, J. C. C., Morabito, J. M. and Lewis, R. K. (1973). *In* "Ion Implantation in Semiconds. and Other Materials" (Ed. Crowder), Plenum p. 87.
Turkevich, A. L., Franzgrote, E. J. and Patterson, J. H. (1967). *Science*, **158**, 636; See also (1975). Adventures in Experimental Physics, 4, 5.
Valković, V., Liebert, R. B., Zabel, T., Larson, H. T., Miljanić, D., Wheeler, R. M. and Phillips, G. C. (1974). *Nucl. Instr. Methods*, **114**, 573.
Valković, V. (1973). *Contemp. Phys.* **14**, 415, *ibid*, 439.
Verona, J., Olness, J. W., Haeberli, W. and Lewis, H. (1959). *Phys. Rev.,* **116**, 1563.
de Waard H. and Feldman, L. C. (1974). *In* "Applcns. of Ion Beams to Metals," (Ed. Picraux *et al.*), Plenum, p. 317.
Wapstra, A. H. and Gove, N. B. (1971). *Nucl. Data Sect.* **A9**, 303 (1972). *ibid*, **A11**, 127.
Weber, R. E. and Peria, W. T. (1967). *J. Appl. Phys.* **38**, 4355.
van der Weg, W. F. and Bierman, D. J. (1969). *Physica*, **44**, 206.
Werner, H. W. (1975). *Surf. Sci.* **47**, 301.
White, C. W. and Tolk, N. H. (1971). *Phys. Rev. Letters,* **26**, 486.
White, C. W., Simms, D. L. and Tolk, N. H. (1972). *Science*, **177**, 481.

White, C. W., Simms, D. L. and Tolk, N. H. (1974). Chap. 23, *In* "Characterization of Solid Surfaces" (Eds Kane and Larrabee), 641.
Whitton, J. L. and Matzke, Hj. (1966). *Can. J. Phys.* **44,** 2905.
Wiedman, L. (1973). Thesis, Brighton Polytechnic.
Williams, J. S. (1975). *Nucl. Instr. and Methods,* **126,** 205.
Wolicki, E. A. (1972). NRL Report 7477.
Ziegler, J. F. and Baglin, J. E. E. (1971). *J. Appl. Phys.* **42,** 2031.
Ziegler, J. F. and Chu, W. K. (1973). *Thin Solid Films*, **19,** 281.

Chapter 9

The Applications of Sputtering

9.1 INTRODUCTION

In this chapter we will demonstrate the very important advantages of using ion beams as a tool for polishing or machining by quoting from a wide range of applications. We have attempted to mention a diversity of fields where sputtering has been used and we are aware that some of the results are rather tentative. However the major aim of this chapter is to stimulate imaginative and more widespread applications of ion beam machining and we will be quite happy to add to the chapter in later editions.

Ion beam machining has, as we have already stated in Chapter 6, the great advantages of being:

(i) a chemically clean process;

(ii) insensitive to the chemistry of the target atoms;

(iii) a machining tool which does not wear or deform;

(iv) accurately controllable in both depth and spatial position to tolerances of 10 to 50 nm;

(v) the sputtering may also be made without the introduction of strain in the work surface.

These qualitative advantages will become more apparent in a quantitative form as we develop our list of examples. We suspect that the list of published results represents only a fraction of the examples considered as many applications are being developed for commercial uses. We shall not consider the economics of ion beam machining except to point out that in the semiconductor industry it is firmly established, for example in the production of beam leads to circuits. The scale of integrated circuitry in microelectronics obviously favours the use of ion beam machining, but even for larger scale items, such as telescope mirrors, the quality of the final surface may outweigh cost disadvantages of ion beams compared with mechanical grinding and polishing. The flexibility of the ion beams in making aspheric systems or devices in chemically "difficult" materials can also be useful.

We will commence by mentioning simple attempts at machining and polishing and progress to more complex structures which may require a combination of techniques.

9.2 MACHINING AND FIGURING

Ion beam sputtering proceeds equally well in all materials, even if they are hard or brittle. A direct application of a collimated ion beam was used by Spencer and Schmidt (1972) to drill small holes (0.3 mm diameter) in diamond. The advantage of the ion machining was that the wall of the hole was left in a polished state and the adjacent regions were free of strain or damage. Their mask defined a circular area but there is no reason why the holes should not have been made with a different geometry.

An equally direct use of sputtering was made by Bayly and Townsend (1972) to produce a large area groove (0.75 × 11 mm) in a sheet of barium alumina silicate glass. The main problem was that the sheet was only 25 microns thick and the groove was to be 15 microns deep with a flat polished base. This was achieved without introducing strain in this fragile specimen.

The small amount of strain associated with ion beam polishing is a distinct improvement on the conventional grinding and lapping techniques which can introduce major strain regions extending several microns below the surface. The ability to polish limited areas of the surface makes ion beam polishing attractive as a method for final figuring as well as for finishing aspherics or removing areas of mechanically strained surface (Scott, 1969; Bayly and Townsend, 1973). One of the first optical applications was made by Narodny and Tarasevich (1967) to generate a paraboloid surface by rotating a spherical disc in front of a small ion beam. The beam was traversed across the radius of the disc and the dwell time varied to produce the required shape. The final paraboloid of f/6 aperture and 10 cm in diameter was carved in a block of pyrex. For economy they mechanically worked the surface to a spherical shape and only used the ion beam to figure the final aspheric. The surface contour was within 0.046 wavelengths of the intended parabola. One may note that the production time was comparable with that for grinding of such a surface. Aspheric lens production has also been reported by Kanekama et al. (1973).

If the figuring process is to be used for correction of roughly shaped surfaces then the error assessments must be made in the target chamber of the ion sputtering system. Schroeder et al. (1966) developed a laser interferometer to satisfy this need. During bombardment there is a temperature rise at the impact area on the surface which distorts the interference pattern but the fringe pattern rapidly returns to a true profile as the surface cools. Later developments by Schroeder *et al.* (1971) used a small ion beam for point by point figuring. Their quoted example was a 5 inch disc finished to within 0.01λ. It should be noted that this finish is better than the normal tolerance requirements of mirror design. The ion beam also enabled them to improve selected areas of the surface without the danger of overcorrection or the introduction of dust or polishing marks.

They point out that in the focus conditions of a very small ion beam it is essential to use a single isotope beam to avoid problems of multiple images from other isotopes. Argon beams seem well suited as they are isotopically pure and perform easily in most ion accelerators. Without further development Schroeder *et al.* (1971) estimated the technique could be scaled up to discs of 3 metres in diameter.

One example of glass machining by ion beam sputtering which could not have been duplicated by conventional means was reported by Karger (1973). The problem was to machine a curved surface in the end of a capillary tube without disturbing the central bore (see Fig. 9.1). Because of the axial

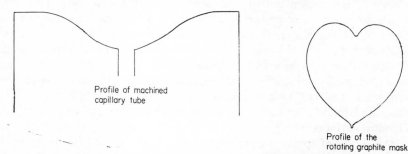

FIG. 9.1. The cross section of a depression which was ion beam sputtered at the end of a capillary tube. Also shown is the cardioid shaped mask which was rotated above the region to produce this change.

symmetry of the problem they chose to use a wide ion beam and rotate a mask above the capillary to determine the amount of sputtering at each point. Clearly the mask design is quite critical and this they computed as a cardioid aperture. The mask was then cut by a pulsed laser tool into a graphite disc.

9.3 POLISHING AND THE REDUCTION OF STRAIN AND SCATTER

Machining over a small depth may equally well be termed polishing. Ion beam polishing forms a useful technique to provide a final polish and remove the scratches, sleeks and surface strains of mechanical lapping. Numerous authors have applied the method to the finishing of optical glasses. Tarasevich (1970) even smoothed ground glass in this way, however the final structure cannot be featureless for a unidirectional ion beam (see Chapter 6). True polishing has been demonstrated by Burdett (1968), Townsend (1970), Wilson (1970) and Kanekama et al. (1973). Pearson and Harsell (1972) deliberately scratched glass surfaces with alumina before polishing with ion beams. Upon rotation of the specimen the surface again became smooth. Topographical features can in general be avoided by rotation or changes in the angle of incidence.

Flatness, or lack of surface features, is only one measure of the quality of a polished surface. Mechanical polishes can produce flat surfaces but this may hide strains or latent cracks which can be revealed by chemical etching or phase contrast microscopy, etc. Bayly (1972) discussed the development of strain cracks and also noted that they could be minimised by sputter removal of the surface (see Section 6.9). Damage regions also scatter light and influence the mobility of surface waves or atomic migration (e.g. in magnetic domains). In lens systems scattered light may lower the imaging efficiency of the system but in laser applications the scattering point may dissipate so much more power than the surroundings that the surface will explode. Surface quality is therefore of major importance for laser rods and optical components used in high power laser systems. Failure in such systems has been reviewed by Glass and Guenther (1973). They quote the surface flaws as being of more consequence than bulk imperfections in the glasses.

Schroeder et al. (1971) measured the fraction of scattered light from glass surfaces and found that ion beam polishing readily equalled the very best mechanical surfaces with some 6×10^{-4} of the light being scattered. However, the choice of sputtering parameters (E, Z_1, θ, T, etc.) is important. If the incident ion is trapped then scattering from bubbles or blisters will result. This can be avoided by lowering the ion energy. They noticed this problem with a Cer-Vit glass (from Owens Illinois Corp.) which retained argon ions. At 150 keV the argon did not polish the Cer-Vit but instead produced a sand-blasted appearance. On lowering the ion energy to 30 keV the problem was resolved. The ion ranges in the two cases are around 205 and 35 nm. In the more open structure of silica the problem did not occur, presumably because the argon was not trapped.

Surface damage by laser beams has been studied by Giuliano (1972). He compared the threshold power required to produce laser damage in conventional and argon ion polished sapphire rods. The threshold value rose in the sputtered rods and the effect appeared to be permanent. This is to be contrasted with the short term improvements achieved by chemical etching. In the Giuliano (1972) experiment the mechanically polished rods had a damage threshold for power levels between 1 and 2·5 GW cm^{-2}. Beyond this, damage was always obtained. Whereas the argon ion polishing raised the threshold level to between 2 and 9 GW cm^{-2} with more than 50% of the test areas being stable beyond 10 GW cm^{-2}. It may well be that the limitation is currently set by inhomogeneities in the bulk crystallinity rather than by the surface scatter.

9.4 SAMPLE PREPARATION FOR ELECTRON MICROSCOPY AND SURFACE ANALYSIS

Ion beam thinning is well suited to the preparation of samples for electron microscopy or surface analysis. The usual advantages are:

  (i) large specimen areas can be prepared;
 (ii) freedom from chemical contamination or preferential chemical attack;
(iii) the technique is viable for essentially all materials;
(iv) the specimen thickness may be reduced *in situ* in the microscope whilst maintaining the sample at a chosen temperature;
 (v) the surface is not mechanically smeared over cracks or other features.

This can be important in fatigued materials where gas inclusions or grain boundaries can be distorted by the strains of mechanical polishing. Whilst we have emphasised that the strain region associated with sputtering is very small it will still be of the order of the collision cascade. This is typically around 10 nm. In electron microscope samples for transmission microscopy this is not trivial compared with the thickness of the specimen, so specimens thinned from one side may bow and contain radiation damage. To avoid the problem of curved samples it is customary to etch the material simultaneously from both sides. The beam profile and the angle of incidence to the sample may be changed by rocking the sample in the beams. This can remove false surface features and offer a range of thicknesses to make the transition from the bulk to the thinned region.

Samples for microscopy are very fragile and rather than etch a thin region from a more rugged block of material many samples are uniformly thinned and then supported on a grid. Ion beam machining can combine the advantages of both methods by sputtering with a mask to leave an integral supporting grid. It was claimed that ion beam thinning was the only possible method of sample preparation for the fragments of friable moon dust studied by Radcliff and Heuer (1970).

Where comparisons of thinning techniques have been made (e.g. Barber, 1970) no obvious differences appear between samples prepared by sputtering or ultramicrotomy except that larger samples are possible with the ion beam approach. Among the earliest samples to be prepared in this way were silicon samples (Heitel and Meyerhoff, 1965) but other specimens include brittle or hard substances such as SiC or Al_2O_3 (e.g. Drum, 1964; Dugdale and Ford, 1966; Bach, 1970), to metallic alloys (e.g. Flewitt et al., 1970, 1972). Many more examples are quoted by Holland (1972).

The Flewitt and Tate (1972) work demonstrates the major advantage of ion beam thinning over chemical etching. These authors prepared a series of molybdenum–ruthenium alloys. They wished to preserve the phase relations and grain structures of the samples in this series and the development of a matching series of reliable chemical etchants would have been an additional major task. The problem was bypassed by sputter etching. A similar advantage was apparent in their study of superconducting alloys.

Surface analysis is simpler, in terms of sample preparation, than electron microscopy because the specimen is normally only thinned on one side and is part of a more substantial item. However, sputter etching is useful in the preparation of the various layers of the surface. Both step and taper profiles have been used with examples ranging from metals, semiconductors, oxides to glass (e.g. Stoddart and Hondros, 1972, Chappel and Stoddart, 1974). The method of analysis can be Auger electron spectroscopy, photoelectron spectra, ion mass spectrometry or ion scattering (see for example Chapter 8) and in these applications ion sputtering may now be accepted as the standard method of sample preparation. A note of caution is that a difference in sputtering rates of the elements of a composite system can lead to errors.

9.5 THE TUNING OF QUARTZ CRYSTAL RESONATORS

The resonant frequency of quartz crystals is a function of their thickness and the mass of the electrodes. Crystal tuning is therefore possible by lapping and polishing. Whilst this is quite suitable at low frequencies one finds that by 20 MHz the samples are so thin and fragile that normal handling is uneconomic. Higher frequencies can still be achieved if they operate in a higher mode than the fundamental but this has the undesirable side effect that the inductance increases, as the square of the harmonic, so the crystals are more heavily damped. Norgate and Hammond (1974) discuss this problem in their review paper on sputtering and in their own laboratory have ion beam thinned resonators to operate at fundamental frequencies up to 50 MHz. Two side benefits of this mode of preparation are good thermal stability and a low series resistance ($<$ 10 ohms).

The metal electrodes should also have a low resistance and add a minimal mass to the quartz. These conflicting requirements can be met by using high conductivity aluminium with the electrode pattern deposited in a groove in the surface of the crystal. Again ion beam sputtering seems an ideal method of machining the groove when the pattern has been defined by a mask. Lukaszek (1970) has made such mask patterns and achieved operating frequencies of 200 MHz in a fifth harmonic mode, without broadening of the resonances.

9.6 PIEZOELECTRIC TRANSDUCERS

A similar application of ion beam thinning has been used for transducers of piezoelectric materials mounted on non-piezoelectric substrates. The technique employed by Beecham (1969) was to deposit a thick ($\approx$ 135 microns) layer of $LiNbO_3$ and ion beam thin this to 2 microns. The resultant device operated at 1·8 GHz. Beecham used a radio frequency sputter source for the deposition of the epitaxial $LiNbO_3$ with an argon–oxygen gas mixture. The oxygen was intended to oxidise the aluminium defining mask and so inhibit it from sputtering. Presumably this was not always successful as he reports that sometimes the surface of the thick layer was conducting but this could be removed by a light chemical etch. Huang *et al.* (1974) have achieved thinner layers of $LiNbO_3$ and formed delay lines which had low loss and wide band performance up to 11 GHz. Their samples at 0·25 microns thick are probably the thinnest that have been made so far.

9.7 HIGH DEFINITION ETCH PATTERNS

The previous examples of sputter machining have been of relatively large scale effects. However, a combination of sophisticated masking techniques, sputtering and ion implantation doping produce the complex structures associated with integrated circuitry. There are of course many examples that one can offer from the semiconductor industry but both the sputtering and masking techniques may be applicable in other fields. Related areas of surface acoustic waves, integrated optics, magnetic bubble memories will be mentioned and the list could be extended with a little imagination.

One of the first semiconductor applications was in the definition of beam leads to circuits. In some devices platinum and gold layers were deposited for different purposes and the chemical similarity of the metals made it impossible to chemically etch a lead pattern in the gold without also attacking the platinum. Ion sputtering avoided the problem. Another simple semiconductor example is resistor value trimming. However, the more elegant applications usually require a masking pattern. The manner in which the mask is made should not influence the final structure and one only has to consider the edge definition provided by the mask and the shape of the step. Ideally one might like to produce a step or a tapered edge. Chemical development of the mask pattern rarely achieves this. If this is a serious problem then an ion beam directed at an angle to the mask could remove the non vertical areas of the masking material (Allison *et al.*, 1969).

Photoresists which use photographic grains are limited by the grain size and for fine details it is often convenient to write the pattern with a finely focussed electron beam. The beam position is programmed. One slight problem is that resist materials are mainly composed of light elements so sputter at a greater rate than the substrate. The solution is to use thick masks or go through a more complex procedure to metallise the masking areas. Some relative sputtering rates of photoresists and commonly used semiconductor materials are shown in Fig. 9.2 for the case of low energy argon ion sputtering.

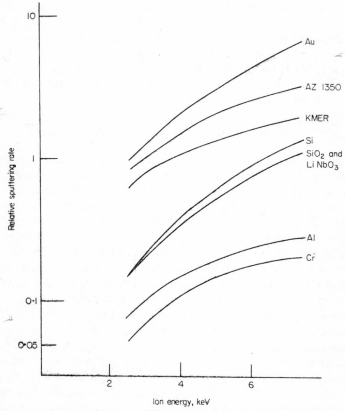

FIG. 9.2. Ion etching rates for a variety of materials which are used in microelectronics. The data is in units of 10^3 nm per hour and was measured with an argon ion current density of $1 \cdot 4$ mA cm^{-2}.

9.7 HIGH DEFINITION ETCH PATTERNS

Pattern resolution was reviewed in 1972 by Garvin and later by Garvin et al. (1973). They discussed masks written by both electron beams and holographic patterns and decided that ion sputtering could be used in either case, without undercutting the mask, up to the resolution set by the masking pattern. This is not true of chemical etching and Fig. 9.3 compares a 10

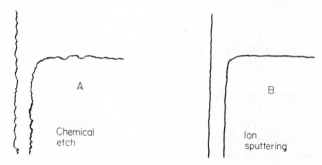

FIG. 9.3. A comparison of the edge definition of a 10 micron wide strip made by (a) chemical and (b) ion sputtering.

micron wide pattern etched into silicon by chemical and ion beam methods. For the same mask quality it is clear that the ion beam avoided problems of sideways attack into the mask region.

A good example of a device which could not have been readily manufactured by other than ion beam sputtering is a Schottky barrier diode which is to operate up to 60 GHz. The geometry of the device, shown in Fig. 9.4,

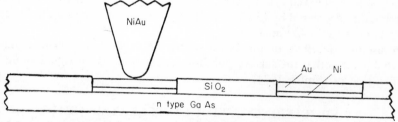

FIG. 9.4. A schematic view of a high frequency Schottky barrier diode made by sputtering. The central contact regions are accurately circular and 2 microns in diameter.

developed by Young and Irvin (1965) and Kahng and Lepsetter (1965), consists of a 2 micron diameter hole etched through a silicon dioxide insulating layer to a thin epitaxial n-type layer. A network of such holes is etched by the ion beam and contact is made by a metal probe through a hole. It is essential for high frequency operation that not only should the hole be this small but also the side walls must be circular and polished, or the fringing capacitance will shunt the diode.

Similar high resolution machining requirements are set for junction field effect transistors (Wolf et al., 1970). In other high resolution semiconductor applications fine grid structures of gold have been machined by Haller et al. (1968). The structure used lines of 0·14 micron width spaced at one micron intervals.

9.8 SURFACE ACOUSTIC WAVES

It is a small step from the semiconductor grid patterns to the ion beam writing of antennae for surface acoustic waves. Acoustic waves are non-dispersive waves which propagate along crystal surfaces. Their velocity is only 10^{-5} the velocity of electromagnetic waves so there is considerable interest to use them in devices for delay lines or pulse compression (e.g. for radar signals). The wave can be monitored at the receiving point on the crystal but it should be noted that because of its confinement to the surface layer it can also be sensed during its progress. To generate or detect such waves it is necessary to write antennae patterns on the surface of the crystal. These may be simple interlaced arrays of rods (interdigital arrays) or patterns with a varying periodicity to achieve pulse compression for a range of signals (Marshall et al., 1973). In essence this is simple but technically the problem is to write an accurate antenna with say 3000 elements at a spacing of 0·7 microns. Garvin (1972) reports the production of such devices as a 1000 to 1 pulse compression delay line on a $LiNbO_3$ surface, tuned to a centre frequency of 300 MHz. An equally complex structure by Wolf (1972) was in the form of a broadband transducer of 244 electrodes, which had a variable spacing that decreased by 15 nm every two wavelengths to give a total change of 0·93 microns in 450 microns.

A useful side effect of ion beam sputtering applied to surface acoustic wave devices is that the surface polish minimises the loss of signal along the surface (see also Smith, 1974).

9.9 SPUTTERING USED IN INTEGRATED OPTICS

A number of potential applications for sputtering and ion implantation for optical frequency components have been proposed but so far there are limited numbers of actual results in the literature which use the high definition capabilities of masked ion beams. An early example (Auton, 1967) was the ingenious production of an infra red polariser by a wire grid of gold on a transparent substrate. Not only can the substrate be prepared with an anti-reflection coating but in addition the polariser is efficient over a broad frequency range. It is of course much more compact than the usual infra red polarisers which have a physically large array of plates set at the Brewster angle. It is also efficient over a wide range of acceptance angles (the Brewster plates are not). Auton's polarizer was a gold grating with a line spacing of 1·1 microns. Pairs of such polarizers gave an extinction ratio for the aligned and crossed configurations of 500:1 at the wavelength of 10·6 microns.

A parallel line structure can act as a diffraction grating (e.g. the X-ray gratings of Franks *et al.*, 1974) and so couple power in or out of an optical waveguide in a surface layer. It is customary to refer to these small scale optical structures of waveguides, couplers, modulators, etc., as integrated optics. This is a little premature as the systems are very simple by comparison with the electronic arrays of integrated circuitry. However, more complex optical circuits consisting of fibre size light sources, lasers, magnetic or electrical plus acoustic and optical components will undoubtedly appear in the future as the individual components have been made. (In the case of fibre size lasers there are reports of diffusion made structures, Saruwatari and Izawa, 1974.)

Diffraction grating masks have been made by both finely focussed electron beams (Garvin *et al.*, 1973) or by a holographic method of crossed coherent beams (Garvin *et al.*, 1973; Shank and Schmidt, 1973; Dalgoutte, 1974). Both approaches are successful and various examples are quoted for gratings in GaAs, ZnS, Si, etc. with periodicities ranging down to 0·28 microns. The electron beam approach is more flexible when the dark to light space ratio of the grating is not unity. The narrowest lines produced by holographic gratings is 0·1 microns. To achieve this a refined technique was used by Shank and Schmidt (1973) as normal interference patterns introduce a minimum periodicity of $\lambda/2$. So, even with a He–Cd laser the holographic limit in free space was 325/2 nm. However, they introduced a prism to compress the pattern.

Blazed gratings can be developed by shaping of the photoresistive mask or by a sputtering beam which is not at normal incidence (Garvin, 1973).

Other simple optical components of light guides, planar lenses and prisms can be machined on the surface in an analogous fashion to chemically etched prisms (e.g. Ulrich and Martin, 1971).

Waveguides may be made by implantation, with a defining mask, or by etching away the surface to leave a ridge which traps the beam of light. If the chosen material is also an electro optic material then a voltage applied to the substrate can modulate the phase of the light beam. A combination of chemical and ion machining techniques was used by Kaminow *et al.* (1974) to produce such a ridge modulator. Simpler structures were reported by Ostrowsky and Dubois (see review paper by Pole *et al.*, 1972).

The existence of a grating over a length of the surface can be used to provide distributed feedback in a thin film laser. The variations in refractive index may be induced by chemical means (Kogelnik and Shank, 1971) or ion sputtering or implantation (Schinke *et al.*, 1972). Laser feedback introduced in this manner produces a compact and stable structure. It is particularly valuable if a broad band dye laser is used for the film material as the grating selectively filters the frequency to tune the laser.

9.10 BIOLOGICAL APPLICATIONS

The sputtering process is a function of the target crystallinity and atomic mass. We mentioned in Chapter 6 that this leads to the development of surface features and so reveals grain boundaries, crystal planes and dislocations. This seems to be an accepted benefit in the use of ion beam thinning in surface analysis or microscopy of inorganic materials. There is no reason why the method cannot be adapted to biological systems and, indeed, Stuart *et al.* (1969, 1970) showed that normal and diseased blood cells developed a different topography as the internal membrane structure of the cell was exposed.

9.11 SPUTTER DEPOSITION

The technique of sputter deposition has a much longer history than sputter machining. Consequently there is a long list of applications and experience with different materials which range from the deposition of single elements to metallic alloys, mixed systems, oxides, nitrides, niobates, etc. The general principles of sputter deposition were outlined in Chapter 6 together with a list of review papers which survey the literature from several viewpoints.

Sputter deposition is commonly used in the formation of superconducting alloys (Johnson and Douglass, 1974) with some 38 different systems being reported by Testardi *et al.* (1974). A notable success occurs in the production of Nb–Ge sputtered films (Testardi *et al.*, 1974) which had a superconducting transition temperature above 23 K. The stringent requirements of superconducting alloys in both the composition and order of the atoms within the lattice determines the conditions for deposition. High energy sputtered particles should be minimised and the substrate temperature during deposition must be correct to achieve both radiation damage annealing and the development of the favourable ordered structure.

One of the few areas where the technique is unlikely to be effective is in the deposition of complex molecules or long chain polymers. Sputtering is essentially an atomic process so the transfer of large molecules by the method is inhibited. However, if organic material has been deposited by sputtering, or reactive sputtering, then it may be possible to form longer chain systems by subsequent polymerisation.

A more complex use of sputter deposition is found in the optical industry where sputter deposition has been applied to the correction of optical surfaces by identifying the low areas in the workface by interferometry. The error map which is then produced is photographically converted to a mask of finely spaced holes in the form of a transmission mask for the sputter deposition of the correcting layer. By positioning the "half-tone" mask of holes a short distance from the workface the correction can be quite uniform. Such a process is economically justified in the production of large scale optical components.

Sputter coatings play a useful role in areas ranging from metallization of ceramics, surface passivation, encapsulation of radioactive material, lubrication, wear protection, surface waveguides to the familiar examples of resistor or component fabrication in the semiconductor industry and the sputter coating of master gramophone records.

9.12 CONCLUSION

In this chapter we have looked at a somewhat random selection of sputtering applications that have been attempted in recent years. The fields of acoustic waves and integrated optics are rapidly expanding and we will undoubtedly see many more applications in the future. In the optical industry the methods of ion beam figuring and polishing have been proved but have yet to be included in production processes. Once this happens we should see many new materials included in optics as the problem of soft or chemically reactive materials is minimised since they can be machined and passivated within the ion beam system. Other materials, such as metallic alloys seem to have been ignored. Whether this is from industrial reserve in publishing initial results or from lack of thought for the potential applications is yet to become apparent.

A general conclusion is that sputter machining is now accepted and it has many advantages for future technology.

REFERENCES

Allison, D. F., Youmans, A. P. and Wong, T. H. (1969). *Electronics*, **42**, 112.
Auton, J. P. (1967). *Appl. Optics,* **6**, 1023.
Bach, H. (1970). *J. Non. Crystalline Solids*, **3**, 1.
Barber, D. J. (1970). *J. Mat. Sci.* **5**, 1.
Bayly, A. R. (1972). *J. Mat. Sci.* **7**, 404.
Bayly, A. R. and Townsend, P. D. (1972). *J. Phys. D.* **5**, L103.
Bayly, A. R. and Townsend, P. D. (1973). *Optics and Laser Tech.*, Oct., 205.
Beecham, D. (1969). *J. Appl. Phys.* **40**, 4357.

Burdett, R. K. (1968). *SIRA Review*, **9**, 156.
Chappel, R. A. and Stoddart, C. T. H. (1974). Cited by Norgate and Hammond.
Dalgoutte, D. G. (1974). *Opt. Comm.* **8**, 124.
Drum, C. M. (1964). AERE Rept. R4563.
Dugdale, R. A. and Ford, S. D. (1966). *Trans. Br. Ceramic Soc.* **65**, 166.
Flewitt, P. E. J. and Tate, A. J. (1972). *Journal Less Common Metals*, **27**, 339.
Flewitt, P. E. J. and Thompson, S. J. (1970). *Metallography*, **3**, 447.
Franks, A., Lindsey, K., Bennett, J. M., Speer, R. J., Turner, D. and Hunt, D. J. (1974). *Phil. Trans. Roy. Soc.* **277**, 503.
Garvin, H. L. (1972). Kodak Microelectronics Seminar, San Diego 1972.
Garvin, H. L. (1973). *Solid State Technology*, Nov., 31.
Garvin, H. L., Garmire, E., Somekh, S., Stoll, H. and Yariv, A. (1973). *Appl. Optics*, **12**, 455.
Giuliano, C. R. (1972). *Appl. Phys. Lett.* **21**, 39.
Glass, A. J. and Guenther, A. H. (1973). *Appl. Optics*, **12**, 637.
Haller, I., Hatzakis, M. and Scrinvasan, R. (1968). *IBM J. Res. and Dev.* **12**, 251.
Heitel, B. and Meyerhoff, K. (1965). *Z. Phys.* **165**, 47.
Holland, L. (1972). *Electronic Components*, **13**, 493.
Huang, H. C., Knox, J. D., Turski, Z., Wargo, R. and Hanak, J. J. (1974). *Appl. Phys. Lett.* **24**, 109.
Johnson, G. R. and Douglass, D. H. (1974). *J. Low Temp. Phys.* **14**, 575.
Kahng, D. and Lepsetter, M. P. (1965). *Bell. Syst. Tech. Jour.* **44**, 1525.
Kaminow, I. P., Ramaswamy, V., Schmidt, R. V. and Turner, E. H. (1974). *Appl. Phys. Lett.* **24**, 622.
Kanekama, N., Taniguchi, N., Watanabe, K., Kondo, M. and Matsumoto, T. (1973). *Sci. papers of Phys. and Chem. Res.* **67**, 25.
Karger, A. M. (1973). *Appl. Optics*, **12**, 451.
Kogelnik, H. and Shank, C. V. (1971). *Appl. Phys. Lett.* **18**, 152.
Lukaszek, T. J. (1970). "Proc. 24th Annual Frequency Control Symposium", New Jersey, US Army Electronics Command, p. 126.
Marshall, F. G., Newton, C. O. and Paige, E. C. S. (1973). *Trans. IEEE on Microwave Theory and Techniques Special Issue on Microwave Acoustics*, MTT21 206.
Narodny, L. H. and Tarasevich, T. M. (1967). *Appl. Optics*, **6**, 2010.
Norgate, P. and Hammond, V. J. (1974). *Phys. in Tech.* **5**, 186.
Pearson, A. D. and Harsell, W. B. (1972). *Mat. Res. Bull.* **7**, 567.
Pole, R. V., Miller, S. E., Harris, H. J. and Tien, P. K. (1972). *Appl. Optics*, **11**, 1675.
Radcliff, S. V. and Heuer, A. (1970). Lunar Science Conference, Houston.
Saruwatari, M. and Izawa, T. (1974). *J. Appl. Phys. Lett.* **24**, 603.
Schinke, D. P., Smith, R. G., Spencer, E. G. and Galvin, M. F. (1972). *Appl. Phys. Lett.* **21**, 494.
Schroeder, J. B., Bashkin, S. and Nester, J. F. (1966). *Appl. Optics*, **5**, 1031.
Schroeder, J. B., Dieselman, H. D. and Douglass, J. W. (1971). *Appl. Optics*, **10**, 295.
Scott, R. M. (1969). "Optical Instruments and Techniques" (Ed. J. H. Dickson), Oriel Press, pp. 63–77.
Shank, C. V. and Schmidt, R. V. (1973). *Appl. Phys. Lett.* **23**, 154.
Smith, H. I. (1974). *Proc. IEEE*, **62**, 1361.
Spencer, E. G. and Schmidt, P. H. (1972). *J. Appl. Phys.* **43**, 2956.
Spencer, E. G. and Schmidt, P. H. (1972). *J. Vac. Sci. and Technol.* **18**, 552.
Stoddart, C. T. H. and Hondros, E. D. (1972). *Nature—Physical Science*, **237**, 90.
Stuart, R. P., Osborn, J. S. and Lewis, S. M. (1969). *Vacuum*, **19**, 503.

REFERENCES

Stuart, R. P., Osborn, J. S., Batsdorf, U. and Ambrose, E. J. (1970). *Nature, Lond.* **227,** 397.
Tarasevich, M. (1970). *Appl. Optics,* **9,** 173.
Testardi, L. R., Wernick, J. H. and Royer, W. A. (1974). *Sol. Stat. Comm.* **15,** 1.
Testardi, L. R., Wernick, J. H., Royer, W. A., Bacon, D. D. and Storm, A. R. (1974). *J. Appl. Phys.* **45,** 446.
Townsend, P. D. (1970). *Optics Technology,* **2,** 65.
Ulrich, R. and Martin, R. J. (1971). *Appl. Optics,* **10,** 2077.
Wilson, R. G. (1970). *Optics Technology,* **2,** 19.
Wolf, E. D. (1972). Proc. 7th Nat. Conf. on Electron Probe Analysis, San Francisco, 24A.
Wolf, E. D., Bauer, L. O., Bower, R. W., Garvin, H. L. and Buckey, C. R., (1970). *IEEE Trans. on Electron Devices,* **ED17,** 446.
Young, D. T. and Irvin, J. C. (1965). *Proc. IEEE,* **53,** 2130.

CHAPTER 10

Applications of Ion Implantation and Ion Beam Analysis

10.1 INTRODUCTION

In this chapter we plan to cover a wide range of examples in which ion beams have been used to modify or study surface properties. Undoubtedly the most familiar examples are in the field of semiconductor devices and the development and acceptance of ion implantation techniques is well documented. Rather than cite examples from this area it is more profitable to indicate the advantages or limitations of ion implantation and then proceed to a wider spectrum of examples in other areas. As with the preceding chapter on sputtering our objective is to demonstrate the scope of the changes which can be made rather than enter into details of a particular problem. Ideally the list of references at the end of this chapter should allow the reader to proceed to more detailed treatments of specific areas, unfortunately this is not always possible as some areas are still in their infancy and a few of the suggested applications have not previously been published. General references also include a mention of conferences which were intended to encourage a diversification in the applications of ion implantation. Not surprisingly the contents of such conference proceedings reveal the potential of the field but progress is mainly in the areas related to semiconductors.

Ion beam techniques can be separated into three distinct areas. The first is the direct use of implantation to make surface devices. The second is to use the flexibility in the choice of ion and substrate to explore or develop a range of new materials. The third role is a diagnostic one in which one can control the impurity content of a sample or study the nature of atomic sites. This is possible by the direct reflection of ions, as in Rutherford backscattering, or by the injection of labelled ions which can reveal the structure around an atomic site by nuclear or electron resonance spectroscopy. All three facets of ion implantation are worth further study and we will mention examples of each.

10.2 SEMICONDUCTORS

Having stated that this area is extensively documented (e.g. an annotated bibliography by Agajanian, 1974) and reviewed (e.g. Mayer *et al.*, 1970;

Dearnaley et al., 1973; Crowder, 1973; Mayer and Marsh 1969; and the general references of Chapter 1); we shall recapitulate why the technique is so successful.

(i) Semiconducting properties are controlled by the position of the Fermi level in the valence–conduction band gap. Major changes in stoichiometry can alter the width of the gap but normally it is simpler to vary the conductivity and charge mobility by controlling the impurity states which fall between the energy bands. To do this may only require impurity levels of a few parts per million and such levels are difficult to produce in a reliable fashion, particularly if there are secondary requirements such as spatial control of the impurities. Diffusion techniques are well developed but implantation is clearly an alternative.

(ii) Electrical circuits may be miniaturized and designed into planar structures so that with a sequence of masking, sputter deposition, diffusion and implantation stages a complete device can be made. The greater the complexity of the circuit, the need for high tolerances, or many separate stages of fabrication if chemical methods are used are all reasons why implantation should be considered. Experience has shown that there is a higher success rate in the production of complex devices made by implantation than by the alternative means; this places an economic advantage on the ion beam methods. The corollary to this argument was that because ion implantation equipment was available it could be used for the manufacture of simple devices, the greater precision and cleanliness of the method being a compensation for minor increases in costs.

(iii) Whilst for some semiconductor applications ion implantation is an alternative approach there are many examples, particularly in high frequency circuits, where it is the only successful method. Generally this is because stray capacitances or inductances are minimised with the higher definition circuitry possible with masks and implants.

(iv) Implants may be made through encapsulating or passivating surface layers. This offers the possibility of thermal treatments during implantation or subsequent annealing of the radiation damage, to relieve strains or precipitate different crystal phases.

This simplified list of reasons why ion implantation has become a standard approach for semiconductor devices should offer some clues to the possible uses in other materials.

10.3 PHOSPHORS

The problem of phosphor design is to choose a material which has a sufficiently wide band gap to transmit the emitted light and to provide energy levels between which there are radiative transitions. An obvious solution for visible phosphors is to choose a wide band gap material and add impurity levels. If the gap is wide then the mode of excitation may be directly from other impurity levels or through states in the normal energy bands. Figure 10.1 outlines a selection of alternative paths for excitation and luminescence. If electrons move to the luminescence centre by thermal or electrical activation then the requirements for suitable states for optical pumping can be removed. In essence the problem of providing impurity (or defect) levels is the same as in semiconductors.

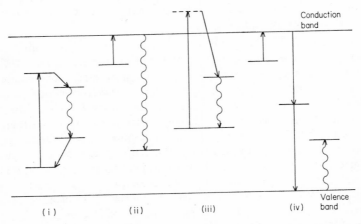

FIG. 10.1. Alternative energy level systems which could produce luminescence. (i) A localised luminescence centre. Excitation followed by relaxation and then luminescence during the decay to the new ground state. (ii) Charge transfer from a trap to a luminescent site. (iii) Transitions involving states in the conduction band and localised levels (e.g. as in ruby). (iv) A process involving charge release from an intermediate centre. (e.g. LiF:Mg:Ti where electron release destroys a defect containing two holes, one of these proceeds to a luminescent site.)

A more complete design specification which includes the response time of the system and overall efficiency is more complex and implies a very complete control of all impurity states since these may influence electron and hole mobilities or alternative decay mechanisms. When one realises that both the added impurities and the "unintentional" impurities are typically measured in parts per million it is not surprising that phosphor preparation is a skilled art.

Variations in the response of a phosphor to trace impurities are readily explored by implantation because one can characterise the phosphor before and after the addition of a known quantity of dopant ions. It is even possible to start with an ultra pure host material and progressively introduce selected impurities. A typical problem which exists, even in established phosphors, is that the addition of one impurity ion is accompanied by a second element which also plays a role in the luminescence process. For example in the field of radiation dosimetry is it customary to use lithium fluoride doped with magnesium and titanium. The addition of optimum quantities of these elements plus a thermal treatment disperses the impurities so that during irradiation electrons are trapped (probably at a lithium vacancy–magnesium defect). In a subsequent heating the thermally released electrons destroy a defect containing two holes, one of which proceeds to a luminescent site (probable a $Ti^{3+} \rightarrow Ti^{4+}$). The mechanism is still in some doubt (e.g. Crittenden et al., 1974) but it is apparent that Ti^{3+} or Ti^{4+} ions are electrostatically unlikely ions to find in a lattice of LiF. It is thought they are stabilised by oxygen ions which were incorporated in the growth process. This is a difficult problem to resolve by chemical means as the oxygen was not intentionally added, it is difficult to detect in analysis and would exist in, at most, a few parts per million. Ion implantation of oxygen or titanium into pure LiF or LiF:Mg offers a means of resolving this problem.

A similar situation occurred in CaO doped with bismuth ions (Hughes and Pells, 1974). The observed luminescence had sharp lines characteristic of gadolinium. This had entered the crystals unintentionally during the doping process. Rather than refine the materials used during the growth stage Hughes and Pells implanted ions of bismuth or gadolinium into pure CaO to establish the origin of the lines.

In addition to the control of emission properties that can be attempted by the choice of dopant this feature may be used in phosphors with spatially varying properties. For example in a scintillation detector the magnitude of the pulse of luminescence can be interpreted as a measure of the range and energy of the incoming particle. There is a limitation that coincident pulses could be analysed as a single high energy pulse. However it is conceivable that a phosphor which changed emission properties with depth could be analysed at several wavelengths and hence resolve this uncertainty. Ion implants would provide a clean method for producing such a system. For example Dearnaley et al. (1971) surveyed the luminescence properties of a range of substrates after implantation with rare earth ions and found a continuous change of emission wavelength with impurity ion. At impurity levels of 1% in alumina or silica substrates they note the cathodoluminescence (or ion excited luminescence) changed from green to yellow with the rare earth series from terbium to dysprosium. The phosphors so prepared were as stable as the original glass and required no annealing to remove radiation damage. This is a desirable property as many of the common phosphors based on ZnS, $CaWO_4$ etc. deteriorate during electron or ion excitation, presumably because the colour centres absorb the light or quench the luminescence process. Such a problem is serious if the phosphor is to have a long operating life without a change in emission characteristics (as is the case with television screens). Ion implanted phosphors may avoid this problem by using ions within the more stable host lattices of silica or alumina or by simplifying the energy level scheme of the luminescence sites, co-activators and traps which occur in chemically prepared phosphors. An example of the latter is the formation of electroluminescent ZnS by implantation of Mn. The implanted material differs from the chemical "equivalent" in that it does not require a co-activator but is still an efficient phosphor (Takagi *et al.*, 1974a).

10.4 CHEMICAL EFFECTS

It is probably appropriate to include a section entitled chemical effects of ion implantation in order to emphasise that one can introduce sufficient ions to produce major changes in chemistry. The point at which one redesignates the surface as a new compound, rather than an impure layer, is somewhat arbitrary. In electrochemical effects such as corrosion, passivation or catalysis the new material may be only a few atomic layers in depth yet able to alter the surface reactions. If thicker layers are involved then both phase changes and new compounds may ensue. Obvious examples are the production of SiO, SiO_2 or Si_3N_4 on silicon substrates or the precipitation and growth of metallic alloy phases, for example in the manufacture of new superconductors.

One frequently finds differences between the bulk compounds and their thin film equivalent where the variations in properties may result from non-stoichiometry, radiation damage or lattice strain, but generally the similarity between the bulk and thin film features is sufficient to identify such features as lattice vibrational absorption bands or the energy band gap. Mixed techniques of thin film preparation and subsequent implantation may offer advantages in terms of cost or final properties (e.g. Stroud, 1972).

Alternative chemical or semiconductor viewpoints may be used to describe ion implantation changes which involve electronic transitions. In the case of photodetectors or nuclear particle detectors where one introduces impurity ions the normal description would be in semiconductor terms. For example the particle detector of Dearnaley *et al.* (1969) which used conventional lithium drifted germanium modified by a gallium implant, or the detectors of Kostka and Kalbitzer (1973) where the rapid changes in impurity level with depth were achieved entirely by implantation into silicon. Similarly one would describe the photodiodes of Marine and Motte (1973) in semiconductor terms because they implanted aluminium impurity ions into Hg Cd Te.

However, the infra red detectors of Fiorito et al. (1973) also utilise Hg Cd Te but here the implant was mercury. Since this is an intrinsic ion of the lattice it might be appropriate to describe the electrical changes in terms of chemical stoichiometry.

10.5 ELECTROCHEMISTRY

Electrochemistry is a field in which ion implantation can play a definite role since one is primarily concerned with surface effects. In addition to the usual implantation advantages of selective implants there may be economic advantages because the implant may involve only a few monolayers whereas conventional techniques may require whole surfaces to be made of expensive materials such as platinum. A recent demonstration of the effectiveness of implantation in electrochemistry was given by Grenness et al. (1974).

The aim of their experiment was to develop efficient and low cost electrodes for fuel cells and electrolysers. The efficiency of such cells is set by a voltage drop which develops at an electrode and due to irreversibility of the electrochemical reaction. This "over-potential" is reduced by using elements of the platinum group. Grenness et al. (1974) attempted to reduce the over-potential by implanting monolayer quantities of platinum into tungsten or tungstic oxide. Their choice of substrates was partly for the convenience of implantation since the masses of platinum and tungsten are similar, and partly because tungstic oxide shows a range of electrical behaviour from semiconducting to metallic. The platinum ions were injected at 200 keV which gives a projected range of 35 nm. The current voltage characteristics of the electrodes were determined with the electrode operating as a cathode. There were some conditioning and hysteresis effects but one can see from Fig. 10.2 that at medium voltages the tungsten and tungstic oxide behave similarly whilst the implanted specimens provided a much higher flow of current. In the example shown the implant was a mere 2×10^{15} platinum ions cm^{-2} (i.e. two monolayers). Electrochemical performance is measured in terms of the log-linear slope, often referred to as the Tafel slope (e.g. Potter, 1956). This dropped to ~ 35 mV/decade for implanted material which is reasonably close to the value of 29.6 mV/decade usually obtained on bright platinum electrodes. The high current flow of 400 μA at the reversible hydrogen potential of -0.242 volts represents the exchange current of the reaction. (The authors only measure the forward current as the evolved hydrogen was removed by flushing with nitrogen.)

The reasons for the success of this experiment are not simply the result of adding platinum to an electrode because the enhanced performance only appeared after a few voltage cycles which "conditioned" the surface and no improvement was introduced into the unoxidised tungsten. The authors

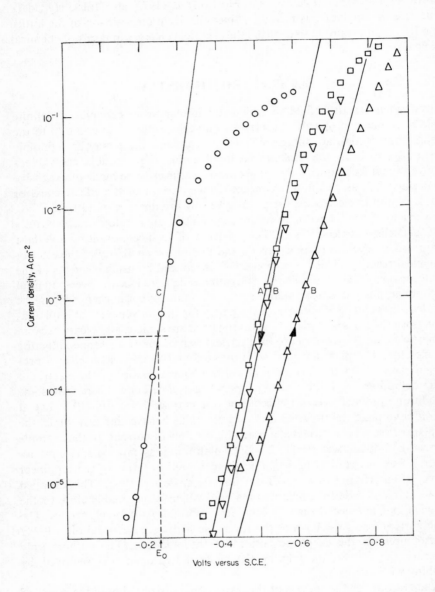

FIG. 10.2. Cathodic evolution of H_2 in a solution of N_2 saturated molar H_2SO_4. E_0 is the reversible hydrogen electrode potential. A Untreated tungsten. B Tungsten anodized at 60 volts (film of 90 nm). Note differences in the direction of the voltage sweep. C Tungsten anodized at 60 volts and implanted with 2×10^{15} Pt ions cm^{-2} (at 200 keV).

suggested alternative mechanisms for the activation by, for example, the dissolution of the oxide above the implant or more likely the formation of a hydrogen–tungsten bronze. For our purposes the detailed model is less important than the fact that electrocatalysis can be induced by minute quantities of implanted ions.

10.6 CORROSION AND OXIDATION STUDIES

Oxidation and corrosion must start at the surface of a metal, either by a direct reaction with exposed grains or by diffusion into cracks and grain boundaries. Therefore, as with wear and friction studies, these properties can be modified by implantation. Again the changes may result from

(i) the modification of the substrate by radiation damage, or

(ii) chemical effects which alter the electrochemistry of the surface reactions.

An example of the first was shown as early as 1961 by Trillat and Hayman who reduced the rate of oxidation of freshly polished uranium by an argon ion implant. Radiation damage does not always lead to passivation and inert gas implants of argon into silicon enhances the oxidation rate (Nomura et al., 1974). One may speculate that these conflicting effects result from the competition of amorphisation of the surface, which in silicon may enhance the etch rate, and the compressive stresses induced by damage which may block the grain boundary diffusion of the oxidant. High dose implants will allow boundaries to move and grain sizes may be more uniform after bombardment. Indeed in oxidation studies of a Zircaloy-4 not only was the oxidation rate suppressed by bombardments with O^+, A^+ or Xe^+ but the oxidation proceeded more uniformly over the surface than in the untreated specimens (Spitznagel et al., 1974). One should expect to achieve a reduction in oxidation rate relatively easily if the surface is granular as it is known (Cox and Pemsler, 1968) that the grain boundary diffusion at low temperatures can be 10^4 times greater than the bulk diffusion. Minor blockages of the grain boundaries by compressive stresses or precipitation of the impurity atoms should lower the total oxidation rate.

Chemical passivation may operate by this means or one may convert the surface to a new alloy or compound which is stable in the atmosphere to which it is exposed. More subtle changes may follow from the electrochemical changes which result from a change in surface potential. An example of indirect chemical conversion is provided by the passivation of silicon surfaces induced by nitrogen implantation. This, according to Fritzsche (1973), forms stable silicon nitride.

The reasoning is far less obvious in the case of passivation, or catalysis, of metal surfaces by metal implantation. An early example was the passivation of copper by boron implantation (Crowder and Tan, 1971). As a first guess as to which ions will provide corrosion inhibition we might choose ions which are already used in bulk metallurgy. Whilst this will not produce a spectacular new material it may allow us to combine properties which were previously mutually exclusive. For example yttrium improves the oxidation resistance of Cr/Ni steels by a factor of 2 or 3 but if it is included in the bulk of the steel there is a problem of intergranular oxidation and the steel is less ductile. To avoid this Antill et al. (1973) implanted yttrium to a concentration of 0.2% within the first 2500 Å of the surface of a steel. This improved both the resistance to oxidation and the adherence of the oxide layer. (Not however to the level obtained with bulk yttrium.) One should note the effect was maintained even for prolonged oxidation.

Conventional alloys are difficult to prepare so the "standard" dopant may not be the only suitable ion and in an impressive demonstration of the rapid survey capabilities of ion implantation the Harwell group (Dearnaley et al., 1973, 1974) measured the oxidation rates of titanium and type 18/8/1 stainless steel after alloying with a range of metal ions. For titanium the effectiveness of the inhibition by the impurity ion correlates with its electronegativity. This is shown in Table 10.1, one also notes that such an ordering is unrelated to ionic size or ionization potential which one might intuitively have felt to be relevant.

TABLE 10.1

Relative effectiveness of ion	Electro-negativity	Position in electrochemical Series	Ionization potential			Ionic crystal radius Å
			1st	2nd	3rd	
Calcium	1.0	$Ca^{2+} - 2e^- - 2.76\,V$	6.1	11.87	52.2	0.99
Europium	1.1	Not quoted	5.67	11.24	$n-q$	0.95
Cerium	1.1	$Ce^{3+} - 3e^- - 2.34V$	5.6	12.3	20.0	1.03
Yttrium	1.3	$Y^{3+} - 3e^- - 2.37\,V$	6.38	12.23	20.5	0.89
Zinc	1.5	$Zn^{2+} - 2e^- - 0.76\,V$	9.39	17.96	37.7	0.74
Aluminium	1.5	$Al^{3+} - 3e^- - 1.706\,V$	5.98	18.82	28.4	0.51
Indium	1.7	$In^{3+} - 3e^- - 0.34V$	5.78	18.86	28.03	0.81
Nickel	1.7	$Ni^{2+} - 2e^- - 0.23\,V$	7.63	18.15	35.16	0.69
Bismuth	1.9	Not quoted	7.29	16.68	25.6	0.96

The Wagner–Hauffe rules (Hauffe, 1965) for anodic oxidation suggest that if the initial film thickness increases as (time)$^{1/2}$ then the process is diffusion limited and the moving entity is either an ion or an electron. The rule seems appropriate for TiO_2 since this is invariably oxygen deficient and so will be an n-type semiconductor with electronic motion controlling the reaction rate. The addition of divalent calcium ions introduces electron traps and so reduces both the conductivity and the rate of oxidation. Changes in electronegativity change this trapping efficiency.

The results on titanium are contrasted by the type 18/8/1 steel. Here the dominant oxide on the steel is Cr_2O_3 which is a metal deficient p-type semiconductor. Metal implantations again show a correlation between the electronegativity and the oxidation rate but in the reverse order to the titanium. In fact all except the bismuth ions enhanced the oxidation rate.

Such experiments have obvious applications and it may ensue that it is even economic to implant expensive metals to provide corrosion resistance since the quantities of ions are minute. However, it is only possible to sense definitive trends if there are no side problems of intergrain diffusion or surface finish. For example the current experiments with implantation into zirconium metal have not led to an obvious correlation with electronegativity, ionic size or other parameter. A recent review of oxidation effects in ion implanted metals has been given by Dearnaley (1974).

10.7 LOCATION OF PASSIVATING IMPLANTS

Not only can implantation passivate surfaces but also the use of ion backscattering or other probe techniques will disclose the location of the implants and their subsequent motion during oxidation. For example the depth analyses of yttrium in stainless steel (Antill et al., 1973; Dearnaley, 1974) by an ion microprobe revealed that the yttrium ions accumulated at the metal–oxide interface (maybe as a perovskite $YCrO_3$). If the layer retreats through the sample as oxidation proceeds then the original implant will be effective even though the oxide has advanced beyond the original implant depth. The yttrium in the stainless steel ($\sim$ 20% Cr, 25% Ni, 1% Nb) persisted as an oxidation and spallation inhibitor as is shown in Fig. 10.3.

Rutherford backscattering was used by Brown and Mackintosh (1973) to follow the movement of ions through the oxide layer on aluminium as the layer thickness was changed by anodisation. They discovered that both the aluminium and oxygen ions are mobile and indeed metal ions injected through the anodic layer (e.g. Ag, Co, Cu, Fe, Ga, Hg, In, Mn, Ni, Tl, Sb, V) are also mobile and pass out through the oxide. The only stable ions appear to be inert gas ions which can be used as markers. They make a cautionary note that without such a marker, whose depth below the surface is independent of whether it was implanted before or after anodisation, it is not possible to make accurate surface analyses by anodic stripping. There are further complications because there are differences between anodic and natural oxides. Implantation of metals through existing oxides results in many metal atoms becoming trapped beneath the metal–oxide interface. However some elements (As, Se, Te) enter into solution in the oxide.

Implantation and analyses such as these allow one to diagnose reactions and atomic diffusion which are quite inaccessible by other methods and clearly reveal the limitations of less direct observations.

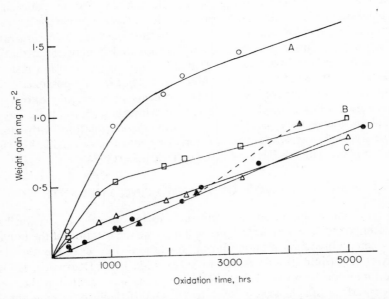

FIG. 10.3. The effect of yttrium implantation on the oxidation of stainless steel, compared with similar steels with yttrium additives. A, 20/25/Nb; B, 20/25/Nb 0·13% Y; C, 20/25/Nb 0·41% Y; D, Y implanted 20/25/Nb.

10.8 POTENTIOSTATIC MEASUREMENTS

Electrochemical measurements may lack the direct lattice information of an ion probe but they still provide data on the state of the surface ions. When a metal is immersed in a corrosive electrolyte, for example zinc in a dilute acid, a reaction such as

$$Zn + 2H^+ \rightarrow Zn^{++} + H_2$$

occurs which can be regarded as an anodic process of oxidation

$$Zn \rightarrow Zn^{++} + 2e^-$$

and a cathodic process of reduction

$$2e^- + 2H^+ \rightarrow H_2$$

The metal takes up a potential, relative to the electrolyte, called the corrosion potential, such that both these reactions occur at the same rate and no charge builds up. The anodic process frequently leads to the thickening of the air formed oxide and thus increasing passivation of the metal as further corrosion is impeded. If the metal is made more positive, anodic polarization, the anodic process is favoured and any oxide layer will be thickened. For the reverse bias, cathodic polarization, reduction of the solution species is favoured. As the metal polarization is varied the current flowing changes and the resulting current density: voltage curves called potentiostatic polarization curves, form the basis of the well established electrochemical potentiostatic polarization method, Uhlig (1971), of testing corrosion resistance. The method has been applied to polycrystalline Cu, Al and Fe implanted with $> 10^{16}$ ions cm^{-2} of 40 keV Ar^+, Al^+, B^+, Fe^+ or Mo^+, Ashworth et al. (1973). In most cases increased passivation seemed to be independent of the implant ion indicating a defect rather than a chemical mechanism. More recent work, Ashworth et al. (1974) on Fe implanted with doses of $> 10^{16}$ ions cm^{-2} of Cr^+, A^+, Fe^+ or Ta^+ at 20 keV showed a doubling of the oxide thickness, mainly by growth of the inner Fe_3O_4 oxide from 13 Å to 33 Å, the outer 11 Å layer of γ-Fe_2O_3 remaining about the same. The change in the potentiostatic polarization curve produced by the chromium implant is shown in Fig. 10.4. The peak current I_p which occurs between -0.4 and -0.8 volts is called the critical current density for passivation and is a measure of the ease of passivation. I_p is smaller for the implanted iron which is hence more passive or resistant to corrosion. The passivation varies linearly with dose and may be compared with the I_p obtained from a range of conventional alloys, Fig. 10.5 where it is seen that 5×10^{16} Cr^+ per cm^2 is equivalent to a 4.2% Cr in Fe alloy and 2×10^{17} Cr^+ ions per cm^2 to a 6.2% alloy.

It is rash to assume that a similar I_p means that the conventional alloy and the implant alloy are identical but the trend of these initial experiments is certainly promising. It is worth repeating that there are problems in preparing a batch of test alloys with a controlled grain size which only vary in the element whose effect we are considering. If one can study such a series by making implant alloys most of the problems disappear. The range of alloys

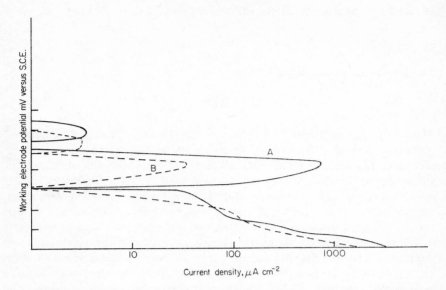

Fig. 10.4. Potentiostatic polarization curves for pure iron (A) and chromium implanted iron (B), measured in a solution of sodium acetate–acetic acid with pH 7·3.

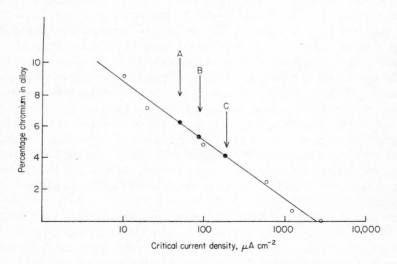

Fig. 10.5. The critical current density for passivation as a function of the chromium content in a Cr–Fe alloy. Also shown are points representing implantation into iron at 20 keV with A, 2×10^{17} Cr ions cm^{-2}; B, 5×10^{16} Ta ions cm^{-2}; C, 5×10^{16} ions cm^{-2}.

could all be made on the same substrate if necessary. It is so easy to do that the whole periodic table becomes available and not just the elements for which a sufficient body of metallurgical knowledge is built up.

10.9 IMPLANTATION OF IMPURITIES FOR LATTICE SITE STUDIES

Research applications of ion implantation include the many surveys of property change produced by particular impurity ions. Frequently it is only necessary to view the results from an empirical basis and decide that a phosphor, semiconductor or metallic alloy has been adequately altered by implantation so no further research study is justified. Alternatively more basic research studies may reveal the way in which the impurity is included into the material and thus indicate why the property changes occurred. If this is possible then we may even predict further changes that are desirable. In addition to this technological role there is a considerable scope for controlled impurity doping to understand the "normal" lattice or defect sites and their interactions. Such studies cross the borderlines between chemistry, solid state or nuclear physics and all disciplines can contribute or benefit from the study of impurity sites.

A major advantage of introducing the impurities by ion implantation is that there is a possibility of doping with a single isotope. This avoids problems of normal chemical abundance and provides control of the nuclear and electronic interactions. Additionally the implanted ion may be a short lived radioactive nucleus. To utilise the information offered by the nuclear interactions one requires resonance experiments such as electron spin resonance (ESR), electron nuclear double resonance (ENDOR), nuclear magnetic resonance (NMR), or measurements of electric quadrupole or nuclear dipole moments. If possible these studies should be coupled to the ion beam analysis techniques of Rutherford backscattering or ion induced nuclear reactions. All these methods are limited to single crystals where the symmetry properties of the lattice and channelling can be used.

The ESR, ENDOR application to insulators or semiconductors is fairly obvious. In defect studies of these materials one looks at the energy level scheme of unpaired electrons or holes which is included by the crystal field and the nuclear spins of atoms involved in, or adjacent to, the defect. The number, magnitude and spacing of the hyperfine lines, together with the defect symmetry, will generally resolve a unique model for the site and the charge coupling. Single isotope induced changes in these patterns obviously simplifies the interpretation of the data.

The use of ion implantation studies in mixed solid state–nuclear physics situations may be less familiar because one is entering several disciplines.

There are also fewer examples of these mixed techniques to be found in the literature. However a convenient starting point is the review paper by de Waard and Feldman (1974) which, together with related papers, appeared in the 1973 conference on "Applications of Ion Beams to Metals". These authors itemize possible changes as:

(i) A Coulomb interaction between the nuclear charge and the surrounding electrons. Because s-type electrons have a finite charge density in the region of the nucleus they perturb the nuclear levels and provide a change in the energy of emitted gamma rays. Mössbauer measurements have the sensitivity to note these shifts in energy.

(ii) There is an interaction between the nuclear magnetic dipole moment and the magnetic fields induced by adjacent electrons.

(iii) Gradients in the electric fields lead to changes in the electric quadrupole moment. These effects become apparent in the measurements with

(a) Mössbauer experiments;
(b) low temperature nuclear orientation;
(c) perturbed angular correlations of emitted gamma rays.

Mössbauer spectroscopy is well developed and so one can analyse the data to give the magnetic dipole or electric quadrupole interactions at the nucleus. This provides a probe into the lattice around the emitting nucleus. Up to the present the use of ion implanted species has been used to study Gd, I, Sb, Tb, Te and Xe ions in iron (de Waard, 1972 and de Waard and Feldman, 1974). It is premature to discuss the lattice sites these reveal as the data is still subject to reinterpretation.

In cases where the quadrupole interaction is negligible the internal magnetic field existing at the nucleus will cause the energy levels of the magnetic dipole moment to split into $(2I + 1)$ hyperfine levels, where the orientation of these states will determine the plane of polarisation of the emitted gamma rays. Under normal conditions thermal energy will equilibrate the energy level populations of the hyperfine states so that the radiation pattern is uniform. However, by cooling to a few hundred millidegrees (~ 0.1 K) the emission pattern can become anisotropic and the ratio of gamma emissions in orthogonal directions, together with their temperature variations, enables one to calculate the hyperfine interactions and hence the lattice locations of the ions. Iron has been the favourite material with implant studies of Au, Ce, I, Sb, Tb and Xe being reported.

Processes (other than thermal) which cause polarized emission from the nucleus may be resolved by the angular correlation between pairs of gamma rays emitted in subsequent steps of the decay process. The directions of

10.9 IMPLANTATION OF IMPURITIES FOR LATTICE SITE STUDIES

emission should be related to the normal states of the nucleus so changes in the correlation pattern are related to the strength of the nuclear coupling to the crystal lattice. If there is a magnetic field perpendicular to the plane of detection then the precessional frequency ($\omega_L = -\mu H/(\hbar I)$) will perturb the angular correlation measurement. Variations on this experiment yield information on the hyperfine interactions, the Larmor precessional frequency and possibly the degree of relaxation of the nuclear states. Interpretation is simple if one finds a single precessional frequency as this implies a single site for the impurity.

A further variation on the experiment is to study recoil implantation by the perturbed angular correlation patterns. Here the excited nuclei recoil into a substrate during the decay. Such studies have mainly been limited to host lattices of iron, nickel and cobalt. Interpretation of this and electric quadrupole data is at an early stage with few experiments indicating a unique defect site. For example in studies of ^{111}In implanted in a range of non-cubic materials (Kaufmann et al., 1974) only implants in Be, Re, Te or Ti gave unique lattice sites. Data from Bi, Hf, Sc, Zn or Zr lattices could not be usefully analysed.

Channelling studies of Rutherford backscattering or induced nuclear reactions clearly supplement the preceding experiments. The impurity is evident, or not, in channelling experiments depending on whether it is substitutional or interstitial. If the latter it will block the channels and so cause flux peaking in the backscattering and high reaction yields for (d, p) or similar events. The limitations of these analyses are:

(i) the minimum impurity concentration;

(ii) changes produced by the experiment;

(iii) problems of interpretation.

To amplify these statements we should note that in practice we may require impurity levels as high as 0·1%. If these are implanted ions they will cause considerable radiation damage and hence the sites may involve strain in the lattice or complex defect sites. For channelling experiments both features are undesirable so annealing may be required. Backscattering measurements with light ions (H or He) are unlikely to produce significant damage but the impurities themselves may cause a change in the long range order of the lattice. The high dopant levels in excess of 0·1% may exceed the solubility limit for the implant or, as in the case of carbon in iron, induce a phase change in the lattice; even though the ion is occupying an interstitial site. Consequently one rarely finds clear cases of only interstitial or only substitutional impurity ions and in many cases the data is such that we must suppose there is a random array of sites. Table 10.2 lists some sites proposed for impurity ions in several metals but one should appreciate that under different conditions more definitive statements may be possible. Also, the use of ion implantation for impurity studies is still in its infancy, as applied to metals, and future results may be more instructive.

TABLE 10.2

Host lattice	Structure	Interstitial	Substitutional	Mixed	Random
Iron	b.c.c.	B, C	Cu, Sb, Te, Au, Tl, Pb, Bi	Ca, Br, I, Xe Tb, Yb	F
Tungsten	b.c.c.	D, ^{3}He, B	Rn		
Niobium	b.c.c.	D			
Nickel	f.c.c.			Bi, Hf, Tl, Pb	B
Copper	f.c.c.	D			B
Aluminium	f.c.c.				B
Gold	f.c.c.				B
Titanium	h.c.p.	O			

(based on de Waard and Feldman, 1974)

10.10 DEPTH PROFILE OF FLUORIDE IN TOOTH ENAMEL

The decay of teeth, which are insulators, is a ubiquitous problem in modern society and the incorporation of fluorine in the tooth enamel has been found to offer enhanced protection. Little is known of the depth profile of the fluorine near the surface of the tooth, without which it is difficult to compare the efficacy of different methods of introducing the fluoride. Mandler et al. (1973) used the $^{19}F(p, \alpha\gamma)^{16}O$ reaction to measure the fluorine profile with a resolution of 0·1 μm. The reaction has a resonance reaction at a proton energy of 672 keV with a cross section of 57 mb. The energy of the protons at a given depth below the surface of the tooth is the initial energy of the proton beam striking the tooth less the energy loss in traversing the calcium hydroxylapatite tooth material to that depth. Thus by increasing the proton energy from 672 keV the energy resonance can be scanned down into the tooth, only the fluorine near a depth where the protons have an energy of 672 keV participating in the resonance reaction. The depth resolution is determined by the width of the resonance (6 keV) and the beam energy spread (3 keV). The depth limit of 2·4 μm is fixed by the onset of the next highest resonance at 835 keV. A high resolution Ge (Li) detector monitored the

6·13 MeV gamma rays from the reaction and was calibrated against CaF_2 or Teflon standards. The fluorine profiles of teeth given three different treatments is shown in Fig. 10.6 where significant differences in both magnitude and shape of the profile are obvious.

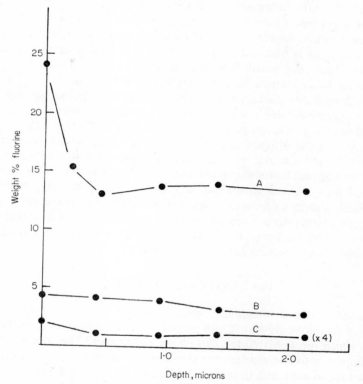

FIG. 10.6. The fluorine concentration profiles of teeth after treatment with: A, a 2·5% solution of zirconyl fluoride for 24 hours; B, a 2·6% solution of sodium fluoride for 24 hours; C, a 1·23% acidulated fluoroplasplate solution for 4 minutes.

Although only three teeth were used in this work and X-rays showed that some changes in the tooth material (and perhaps the stopping power) occurred with the zirconyl fluoride, it is an interesting example of the application of energetic particle beams to the study of biological material.

10.11 ANALYSIS OF VARNISH LAYERS OF VIOLINS

Because violin making is a highly developed art which developed to a peak at a time when modern scientific aids were not available, a great deal of mysticism surrounds the skill of the early violin makers. For example people have speculated on the role of the varnish on both the tone and appearance of the instruments. Modern opinions minimise the effect on the tone, unless there is an excessive thickness, but there is still a question of whether the lustre of the early instruments is a result of many layers of identical or different varnishes. Chemical analyses of varnish scrapings only give the total composition whereas one would like an analysis through the layers. Such an analysis is reported by Tove and Chu (1974) by using Rutherford backscattering. They noted differences in the composition of the varnishes of different makers and also some depth dependence through the varnish layer. It is probably premature to decide between the alternative reasons of wood preparation, deliberate changes in varnish or results of external polishes but the results do throw new light onto a very old problem.

The deliberate layering of varnish on an instrument achieved the result that the refractive index of the final varnish is discontinuous, this produces interference and optical scattering and so enhances the appearance of the surface. We shall now make a chronologically large jump from the problems of acoustics, vibrating surfaces and light scattering faced by Stradivarius in the eighteenth century to a twentieth century technology which is still in its infancy—namely the interaction of acoustic waves and light scattering in thin films which have variations in the refractive index!

10.12 INTEGRATED OPTICS

We have mentioned integrated optics at several points in this book and indicated that not only may it be a growth area in future technology but also one in which ion implantation can play a major part. The concept of integrated optics is to miniaturize optical systems to the point that they can form a planar technology and then interlace them with the developments of fibre optics, electric, magnetic and acoustic wave effects. There is no reason why all the familiar optical components of light sources, lasers, mirrors, prisms, gratings and lenses should not be reduced to a planar system and certainly the efficacy of sputter deposited and sputter machined layers for most of these items has already been demonstrated (Shubert and Harris, 1971; Ulrich and Martin, 1971; Kompfner, 1972.)

If such optical circuits are to become accepted then the problem is to design the other elements of the system such as guide regions, bends, directional couplers, beam splitters, mode converters, filters and phase shifters. Individually all these devices have been fabricated though not always by ion implantation. Previous experience would suggest that if diffusion or sputter deposition is adequate for the development of the layers then ion implantation should be a valid alternative since the depth scale of the guides is typically in the order of a micron (i.e. comparable with the wavelength used) so can be influenced by ion beams of moderate energy and mass for most target materials. The advantage of the ion beam approach is immediately obvious from the calculations of radiation losses at bends or irregularities in the walls of the guides (Marcatili, 1969; Marcuse, 1969). The wall definition should be as high as possible, for example, with r.m.s. variations of less than 10 nm, and here of course implantation is superior to diffusion limited guides. Sputter deposited guides may form epitaxial grains, which in ZnO can be around 0·5 microns diameter, that lead to high scattering losses of 60 dB cm^{-1} in the guide. Such a high loss guide would not be of practical use and an acceptable loss is a few dB cm^{-1}. Implanted guides are less likely to form grains, need not vary in thickness and are safely buried to avoid surface scratches and scatter. One also avoids the problems of surface cleanliness during deposition and subsequent handling. A disadvantage of implantation is that some systems are amorphous (unless annealed). These may guide but lack the lattice properties of electro-optic or magneto-optic crystals.

Many of the early papers describing the theory and demonstrations of integrated optic components have been reprinted in a book edited by Marcuse (1972). We will now mention a selection of these articles together with more recent ones to demonstrate the state of the art (see also Balkanski and Lallemand, 1973, and the special issue of the I.E.E.E., 1975).

Optical waveguides are the simplest structures to have been considered. These are introduced by making a region of the material with a higher refractive index than the surrounds. If the guide cross section is comparable with the wavelength then one requires an index change of around 1% to support at least one mode. Higher changes or wider guides can support more

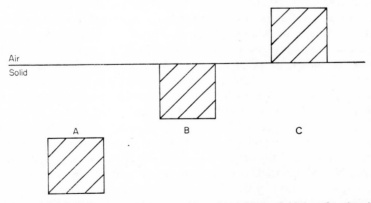

FIG. 10.7. Three examples of optical waveguides with region of high refractive index: A, within the solid; B, at the surface; C, projecting above the surface.

modes. Three alternative geometries are shown in Fig. 10.7. Guide A is a buried guide which is not influenced by the surface; B is at the surface (i.e. as are deposited or diffused guides); C is a ridge waveguide above the surface.

Sputter deposited guides, of type B, have been prepared for many layer-substrate combinations with BaO, ZnO, GaAs, ZnS, LiNbO$_3$, LiTaO$_3$, Ta$_2$O$_5$, silicate glasses and SiON receiving particular attention. Generally the energy loss along the guide is at least 1 dB cm^{-1} (e.g. Corning 7059 glass used by Goell and Standley, 1969) but lower losses of less than 0·04 dB cm^{-1} were reported in organosilicon films (Tien et al., 1972b). As an alternative to sputter deposition layers have been formed by diffusion, both by addition of material (in-diffusion) for (e.g. Taylor et al., 1972, or Martin and Hall, 1972) or preferential loss of one element from a compound (out-diffusion) (e.g. Kaminow and Carruthers, 1973) or ion exchange (French and Pearson, 1970). The manner in which the refractive index is changed is irrelevant so long as the final guide does not introduce too much loss. However, diffusion methods lack the control in the profile of the change which can be introduced by implantation (e.g. Bayly, 1973; Bayly and Townsend, 1973).

Buried waveguides were first made by Schineller et al. (1968) by proton bombardment of silica. Protons produce both ionisation and displacement changes in the solid which contribute differently to the index change. Both produce an increase in refractive index but have different saturation characteristics with implantation dose (Presby and Brown, 1974). Some differences occur in the thermal stability of the two regions but in general no major changes occur below 500°C. Silica has also been used to form waveguides by implanting lithium ions (Goell et al., 1972), lithium and argon ions (Wei et al., 1973, Standley et al., 1972), protons, helium, argon, lithium, bismuth, nitrogen and neon (Webb et al., 1967). It is not altogether clear whether the increase in refractive index arises from compaction of the silica from radiation damage or whether there is also a contribution from chemical effects. Potentially high indices are possible in silica implanted with nitrogen because the index for silica (n_{SiO_2} = 1·46) but both silicon oxynitride or silicon nitride were considered as a vapour deposited layer by Rand and Standley (1972) and layers with an index between 1·48 and 1·54 were deposited which had low losses of less than 0·4 dB cm^{-1}. Because of this nitrogen implants were attempted by Webb et al., 1967, and excellent guide effects were noted for several modes which suggested a significant rise in refractive index, in excess of the 1% change characteristic of the density change from radiation damage. Not all materials increase their refractive index on implantation and a major reduction by 10% is found in LiNbO$_3$ (Wei et al., 1974).

Gallium arsenide has also been used in successful implantation experiments and guides were formed by proton injection (Garmire et al., 1972, Garvin

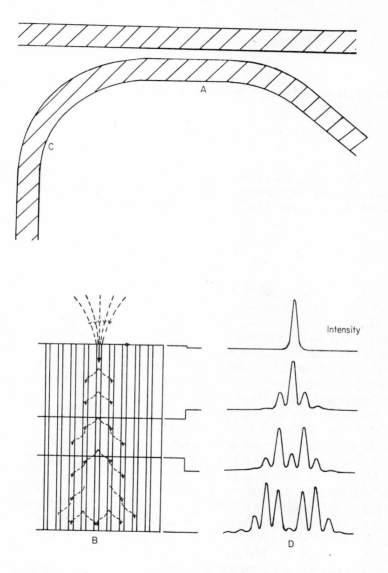

FIG. 10.8. Regions A and B allow coupling between optical waveguides via an evanescent wave. Region C demonstrates how the guides separate with a bend and the results of D demonstrate an actual pattern of energy transfer among a set of parallel guides (after Somekh *et al.*, 1973).

et al., 1973). In this example a single bombardment energy was sufficient to develop a guiding layer because of charge compensation in the n-type GaAs. The condition for the change in the conductivity and dielectric constant is that

$$\Delta\varepsilon > \varepsilon_0 \left(\frac{\lambda_0}{4t}\right)^2$$

where λ_0 is the free space wavelength and t is the guide thickness. Also

$$\Delta\varepsilon = \frac{(N_s - N_c)e^2}{m^* \omega^2}$$

where N_s and N_c are the electron densities in the substrate and implanted layer, e, m^* are the electronic charge and effective mass, and c_0 is the velocity of light in free space. This leads to a condition for light guiding of

$$N_s - N_c > \frac{\pi^2 c_0^2 m^* \varepsilon_0}{4e^2 t^2}.$$

An expression independent of the guided wavelength. The sudden change in dielectric constant implies that a diode junction can also act as a waveguide. If electronic transitions at the junction emit photons then a diode can act a light source and a confining region for the light (i.e. the situation of a laser diode). In addition to diode light sources light from lasers may be introduced into the guide or laser systems may be an integral part of the guide. Fibre optic size lasers have been made by diffusion (Saruwatari and Izawa, 1974) so implantation methods should succeed equally well.

In Chapter 9 we mentioned the use of sputtering or implantation to develop periodic changes in the surface structure for diffraction gratings. These may both couple light in or out of the guide or act as a wavelength filter for laser action in the guide (an advantage in externally pumped dye lasers).

If more than one waveguide is to be used on a surface then it is necessary to bend the guide and make coupling between different guides. In the original scheme of Schineller et al. (1968) it was intended that two guides should run parallel and couple via the evanescent wave, Fig. 10.8(a), this effect was demonstrated by Somekh et al. (1973) who made parallel guides in GaAs by a proton implantation. The guides couple in a directional fashion so that in an array of parallel guides the energy is transferred sideways between the guides in the forward direction Fig. 10.8(b). Separation of the guides with curved sections Fig. 10.8(c), has also been demonstrated (e.g. Goell, 1973). Other guide geometries could provide frequency filtering, for example in the set of guides regions shown in Fig. 10.9 (Marcatili, 1969) the central loop or disc is resonant to frequency f_2 so only this frequency is coupled between the two guides.

Phase and mode control of the propagating wave have been attempted by a variety of methods. The ridge waveguide (Kaminow *et al.*, 1974) on a LiNbO$_3$ film shifted the phase of the light by a voltage applied to the substrate. LiNbO$_3$ was used in electro–optic modulators of Channin (1971) and Kaminow *et al.* (1973), Schmidt *et al.* (1973). The relationship between

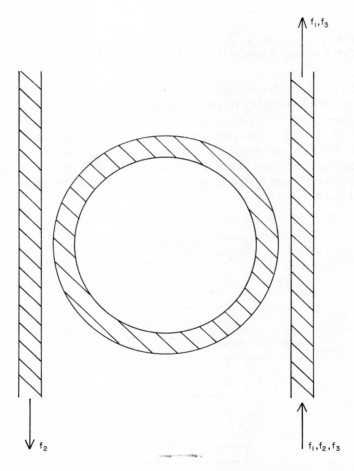

FIG. 10.9. A frequency filter comprising a pair of waveguides coupled by a loop which is resonant at frequency f_2.

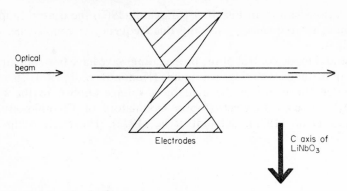

Fig. 10.10. A thin film electro-optic phase modulator. The electric field is applied across the surface of the crystal, parallel to the c axis, and perpendicular to the guided light beam.

waveguide, electrodes and crystal axes are sketched in Fig. 10.10. Other schemes include Faraday rotation of the plane of polarization of the beam in a magneto–optic crystal (Tien *et al.*, 1972a) and acoustic wave interactions (Kuhn *et al.*, 1970). An experimental electrode arrangement is shown in Fig. 10.11. The surface acoustic wave generates a strain pattern which diffracts the light beam. All three methods may be developed in the future but at present none of the methods achieves a complete switching of the optical power.

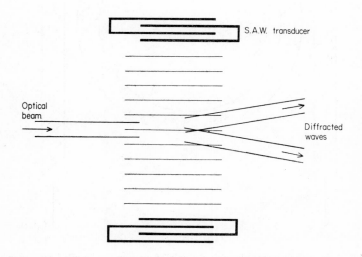

Fig. 10.11. Diffraction of an optical beam by an acoustic surface wave.

Finally one should consider the power levels existing in an optical waveguide. The guide dimensions are typically one or two microns and if milliwatts of laser light are introduced then the power level can reach a megawatt per square centimetre. At such levels one must seriously consider the possibility of non linear effects and second harmonic generation. The reverse view is that waveguides form an ideal situation for production of the second harmonic (Tien, 1971). In bulk systems one uses a suitably oriented birefringent crystal to match the phase velocities of the fundamental and harmonic wave.

In a thin film the phase conditions are set by the choice of mode and the thickness of the guide rather than the dispersion curve. The advantage of using thin films is that many materials have large non linear optical coefficients (e.g. GaAs, GaP, ZnS, ZnTe) but are not birefringent so are useful in a guide but unsuited for bulk crystal generation of harmonics. There are obvious variations on this theme for example to make a film with a high non-linear coefficient of polarization on the top of a birefringent crystal (Tien *et al.*, 1970) or to vary the thickness of the guide (Anderson and Boyd, 1971). Alternatively one may be able to alter the dispersion curve for the film by implantation.

The present picture of integrated optics is that the constituent elements exist and when ultra low loss optical fibres (i.e. a few dB km^{-1}) become commonplace then interest in this field will rapidly expand. This expansion will undoubtedly involve the techniques of ion implantation.

10.13 FRICTION AND WEAR

An enormous range of applications for ion beam technology exists in the processing of metal surfaces and it is surprising that this area of science has attracted so little research. The benefits accrued from such research can be applied to all three of the interrelated areas of friction, wear and corrosion of surfaces. Indeed the conventional treatments of metallic surfaces by procedures such as case hardening, alloying, plating, plasma spraying, painting or sputter coating have already identified many of the problems and also indicated that the critical surface regions are shallow enough to be formed by ion implantation. The concentration of surface ions may represent a few per cent of the atomic concentration in the outer layer and this will involve longer implantation stages than are customary in the semiconductor industry but nonetheless the economic arguments may still justify this approach. The potential market could be very large, particularly where friction and wear problems are severe at limited regions of the surface.

We shall first consider the problems of friction and wear which occur between two metal surfaces. The conventional attempt to lower the frictional losses and reduce wear is to introduce a lubricant between the moving metal faces. This is never completely successful and in many instances the lubricants are expelled from the region where they are most needed so the problem returns to one of metallic friction. It is also important to note that the lubricant can cause chemical changes on the surfaces. In all frictional problems the coefficient of friction will depend on the strength of the junction material. This is equally true for liquid or solid lubricants, interfacial oxides or adhesion between the metal surfaces. The more obvious problems include the formation of microwelds between projecting blocks of the materials, the size and nature of the protrusions, for example whether they are an integral part of the surface or a result of previous grinding action; the grain size, dislocation density and presence of microcracks. All these features can be altered by the surface preparation and subsequent thermal or chemical treatments. It is simple to generalise in this way and far more difficult to predict the effect of a particular change. For example an oxidising process may form a surface layer which is only weakly bonded so that it abrades during wear but because it is being continuously formed it also preserves the original metal face. This advantageous use of chemical attack may be offset by an etching of grain boundaries which will then lead to major particles being torn from the surface (Archard, 1953).

Competing effects may provide results which *are* time dependent or a function of the contact force between the faces and the number of times that the surfaces slide across one another. Temperature changes induced by the frictional forces can lead to various chemical reactions with the surrounding atmosphere and one may form either lubricating oxide layers (e.g. Wakelin, 1974) or abrasive particles (such as Fe_3O_4) on the metal surfaces.

One method of burying the strained layer which is formed during machining is to sputter deposit an overlay which acts as a solid lubricant. The usual advantages of sputter deposition also occur here so the surface is both cleaned and covered with atoms that are well bonded to the substrate. The bond strength is equally good for rough or polished substrates and the choice of film need not be limited to the known solid lubricants (e.g. Au, Ag, MoS_2, WS_2 etc.) because the adhesion characteristics of sputter coatings are generally good. The amorphous nature of this continuous deposit was demonstrated by electron microscopy studies of layers of MoS_2 (e.g. Spalvins, 1971). He found that the MoS_2 layers were amorphous or, at worst, had an average particle size less than 30 Å. These films were between 2000 and 6500 Å thick and well bonded so that they did not peel even when the substrate rods of nickel or inconel were fractured. To minimise problems of oxidation Spalvins conducted friction tests in a high vacuum of 10^{-9} Torr. The measured coefficient of friction was a function of the load, the velocity of the probe across the metal face and it also changed with repeated measurement. At low loads the coefficient was 0·06 but this value reduced to between 0·03 and 0·04 during the first few hundred transits. At high loads the initial values was lower, at 0·02, but steadily rose during the experiments until the test failed after some 38,000 cycles.

The success of sputter deposition even for films as thin as a few thousand Ångstroms suggests that ion implantation should be equally successful. Implantation ensures atomic dispersal of the lubricant ion and strong bonds to the substrate. However the act of implantation also increases the atomic density of the surface layer by inducing radiation damage and adding further ions. Both effects produce compaction of the surface layer. This automatically "case hardens" the surface, closes the microcracks and locks in the surface grains. The compaction probably reduces the number of surface, or crack, sites which are suitable for chemical attack so should additionally provide some corrosion resistance. (Indeed it does—see Section 10.6.)

One may also speculate that in some cases the diffusion of the implanted ions into the solid will proceed as the surface layer is worn away and the reduction in wear and friction will be maintained to depths beyond the range of the original implanted ions.

In a series of experiments Hartley *et al.* (1973), measured friction, wear and chemical changes as a loaded tungsten carbide ball is repeatedly driven across an implanted surface. They found that many implanted metals produced a reduction in the coefficient of friction of the steel surface and such changes were apparent even with the addition of "soft" metals like Ag, Sn or In. Table 10.3 summarises some of the changes that they observed.

TABLE 10.3 Friction Data for Ion Implanted En 352 Steel Tested in Air

Ion	Dose (ions cm^{-2})	μ_{En352}	μ_{impl}	Remarks
Kr	$2·8 \times 10^{16}$	0·24	0·24	Friction peak at implantation boundary
Sn	$2·8 \times 10^{16}$	0·24	0·09	
In	$2·8 \times 10^{16}$	0·30	$0·31 \pm 0·05$	Erratic; transfer of ions during wear
Ag	$2·8 \times 10^{16}$	0·22	0·26	Erratic; adhesion
Pb	$6·3 \times 10^{16}$	0·23	$0·33 \pm 0·08$	Stick-slip; adhesion
Mo	$2·8 \times 10^{16}$	0·26	0·24	
S	$6·1 \times 10^{16}$	0·20	0·19	
Mo + 2S	$2·8 \times 10^{16}$ + $5·6 \times 10^{16}$	0·26	0·20	

It is apparent from the table and Fig. 10.12, that not all the ions produced a reduction in the coefficient of friction. Inert gas ions of krypton produced no change and lead introduced a complex stick-slip action on the surface which increased the coefficient of friction. The lead was sufficiently weakly bonded that it sheared and steps in the surface were detectable by subsequent microscopy. The transfer of lead to the surrounding, non-implanted, steel produced an increase in friction on the untreated steel which changed with the number of passages of the measuring ball. Hartley *et al.* (1973) speculate that the characteristics of the two regions are caused by the formation of soft lead oxide and abrasive iron oxide particles.

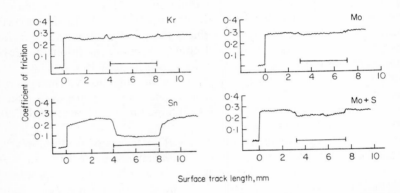

FIG. 10.12. Frictional changes induced in steel as a result of ion implantation. The markers indicate the position of implant. The curves are for Kr^+ at 400 keV, $2 \cdot 8 \times 10^{16}$ ions cm^{-2}; Mo^+ at 400 keV, $2 \cdot 8 \times 10^{16}$ ions cm^{-2}; Mo^+ 400 keV ($2 \cdot 8 \times 10^{16}$ ions cm^{-2}) plus S^+ at 150 keV ($5 \cdot 6 \times 10^{16}$ ions cm^{-2}); and Sn^+ at 380 keV, $2 \cdot 8 \times 10^{16}$ ions cm^{-2}.

No such problems occurred with molybdenum or sulphur implants but it is interesting to note that consecutive implants of Mo and S in an atomic ratio 1:2 produced a greater reduction in the friction than either ion did when used alone.

These experiments have involved a number of ions implanted at between 150 and 400 keV and the results have been encouraging for both steel and copper substrates. Similar experiments by Takagi *et al.* (1974b) have shown that small doses of boron ions (10^{15} ions cm^{-2}) significantly increase the hardness of steel and reduce abrasion by a factor of 3. Undoubtedly similar experiments will be attempted with a wider range of ions into metals or other substances.

Hartley *et al.* (1973) also report some preliminary measurements of the reduction in wear of surfaces as a result of ion implantation. To quantify the

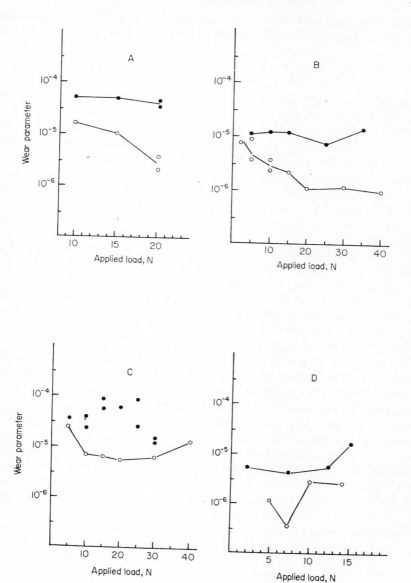

Fig. 10.13. Changes in wear parameter for various systems ●—● represents unimplanted material and ○—○ is after implant. Results for: A, Mo$^+$, 400 keV, 2.8×10^{16} ions cm^{-2}, 440C steel disc, 440C steel pin; B, N$^+$, 35 keV, 10^{18} ion cm^{-2}, mild steel disc, 440C steel pin; C, B$^+$, 40 keV, 5.6×10^{16} ions cm^{-2}, mild steel disc, 440C steel pin; D, B$^+$, 40 keV, 10^{17} ions cm^{-2}, copper disc, carbon pin.

changes they used a metal or carbon pin to bear on an implanted rotating disc to form a groove. They then define a wear parameter K as

$$K = 3 \times \frac{\text{(volume change)} \times \text{hardness}}{\text{(load)} \times \text{sliding distance}};$$

(K has the advantage that it is independent of the speed of rotation of the disc). In situations of severe wear particles are torn from the surface and K is about 10^{-2} or 10^{-3}. However in systems which are protected by self generated oxide films K is reduced to 10^{-5} or 10^{-6}. In this initial survey the Harwell group implanted N^+, B^+ and Mo^+ ions at 35, 40 and 400 keV respectively into materials ranging in hardness from copper, mild steel to 440 C steel. In all cases there were improvements in the wear parameter with typical changes of a factor of 10 at high loads. Figures 10.13 (a), (b) and (c) demonstrate this for examples of N^+ in mild steel, Mo^+ in a 440 C steel and B^+ in copper.

As one might expect the improvement is a function of the implantation dose and these examples involve some 10^{16} or 10^{17} ions cm^{-2}. Not all elements alloy into the metals and for some ions these implants exceed the solubility limits of the substrate. For metallic implants this may only lead to precipitation. Nitrogen implants show a maximum in the curve of wear improvement versus ion dose, which correlates with the maximum stress in the surface (Hartley, 1976).

The mechanisms which induce hardening are unknown and one can speculate that they are a mixture of changes in grain size, compaction or chemical effects. For example the addition of boron or nitrogen ions in steel accelerates the rate at which oxide layers are formed. Alternatively nitrogen can enter the steel as a hard nitride. There is some evidence for chemical changes as a result of implantation as measured by the reduction in iron oxide debris that accumulates in wear tests of steel. Hartley (1974) has also measured the compressive stress produced by argon ion implants in steel and shown that it approaches several thousand p.s.i. after large dose implantation. (He used the method of the deflection of a cantilever which was suggested by Eer Nisse in 1971.) For the purposes of this chapter the mechanisms are far less important than the fact that ion implantation can reduce wear of both soft and hard materials by an order of magnitude.

10.14 SUPERCONDUCTIVITY

Research into superconducting alloys has been very intensive and is based on a good theoretical understanding of the process. However, the development of new alloy phases can be a difficult and time consuming task; there are also many problems of solubility and phase stability (e.g. Fischer, 1970). In such a situation ion implantation can provide a useful alternative path for the production of test specimens and the chemical purity of the method plus the control in the distribution of the implant avoids many of the metallurgical problems in alloy production. Whilst a range of alloy concentrations may be prepared it should be noted that implantation times also increase with dose and the method might not be appropriate for commercial production of the final alloy. The problems of phase precipitation and grain size are minimised by implanting into cooled targets and many superconductor workers have used targets at less than 10 K. Subsequent heat treatments can then be used to modify the alloy.

The search for new materials is in part to test the theoretical predictions and in part to increase the performance of existing materials as measured by the values of critical field, critical current density and critical temperature. (An example of a good superconductor, Nb_3Sn, has values such as $H_c >$ 200 kOe; $J_c > 10^6$ A cm^{-2} and $T_c \sim 18 \cdot 5$ K.) The measured performance of the material is frequently limited by weak sections or the interaction of flux vortices with the surface. Pinning of these vortices also produces higher performance characteristics which are more representative of bulk properties.

The separation of surface and bulk effects has been attempted by Bett and Howlett (1970), Chang and Rose-Innes (1970) and Freyhardt et al. (1974). Chang and Rose-Innes implanted a strip of molybdenum into the side of a cylinder of very pure Ni–Mo alloy and then measured the current flow as the specimen was rotated in a magnetic field parallel to the cylinder axis.

The molybdenum concentration in the strip was increased by 3% during implantation. This asymmetry in the cylinder induced a critical current variation as the cylinder rotated in the magnetic field. The asymmetry was different for rotation of the strip by 180° from which it was possible to deduce the direction of the Lorentz force on the fluxons. Freyhardt et al. (1974) also introduced an anisotropy into the critical current by pinning the fluxons in the surface of a type II superconductor with voids. They generated voids by irradiation with 3·5 MeV nickel ions into niobium foils containing oxygen. To develop the high vacancy concentration required for void formation the implantation was made at 900°C. The resultant void structure contained a mixture of small scale voids (~ 30Å diameter) and large voids (~ 280Å diameter). An interaction between fluxons and the larger voids altered the dependence of critical current with magnetic field and increased the critical value of the current.

Alloy preparation for superconducting materials has ranged from the formation of conventional alloys (for example tin implants into niobium, Bett and Howlett, 1970) to the exploration of new materials in regions of the periodic table which had not shown superconductivity. Buckel and Stritzker

(1974) formed palladium alloys which were non magnetic by the addition of hydrogen atoms. They argue that the hydrogen reduces the number of free electrons by bonding with the electrons of the d band. It is difficult to introduce high levels of hydrogen by chemical means as it diffuses out from the metal except at very low temperatures. A compromise approach was to introduce hydrogen chemically and then add further protons with the target cooled to 10 K. Not only does the mixture show a superconducting transition (at 9 K) but it is a simple step to vary the isotope and deuterium implants produced a higher critical temperature of 11 K. Buckel and Stritzker (1974) then surveyed a range of other alloys using He, B and Li implants. The trend of a higher critical temperature with higher mass implant led them to a (Pd–Ag–Al) alloy with $T_c \sim 10$ K. Additions of hydrogen raised the value to 16 K. This work is interesting because one is using elements that were previously thought to be unsuited for superconductivity and yet the alloys show high critical temperatures. Other authors who have used ion implantation to survey the range of materials are Buckel and Heim (1974) and Meyer et al. (1974). The latter group introduced ions into Ti, Zr, V, Nb, Ta, Mo, W, Re, Nb_3Sn, NbC and NbN. In general they found radiation damage and inert ions lowered the critical temperature but chemical additions without radiation damage (i.e. after annealing) could increase the performance of the superconductors.

This is clearly a field where ion implantation will be used more extensively, both for new materials and for examining the role of impurity ions in established alloys.

10.15 ION IMPLANTATION AND MAGNETIC BUBBLES

The search for systems which could provide a high density information store with easy access has led to the study of magnetic bubbles in garnet and orthoferrites layers (e.g. Bobeck, 1967; Brown, 1971; Copeland, 1971; Hayaski et al., 1972).

The layers are usually deposited epitaxially and one finds that if there are constraints on the layer, from an external field or a mismatch between the layer and substrate crystal, then the film forms in a particular plane of magnetisation and this can be normal to the film. The whole film is not identically oriented so regions will be polarized up or down with respect to the film. The magnetic domains are separated by Bloch walls where the spin alignments reverse. On deposition these may be in a random pattern but if one applies an external magnetic field the opposed domains will shrink and finally leave cylindrical regions called magnetic bubbles. Control of the movement and polarization of the bubbles forms the basis for an information store.

Since the process involves a planar technology of thin films one might expect that ion implantation could be used to advantage in the definition of suitable systems. We may therefore list the problems of bubble formation.

(i) The magnetisation of the layer is a function of the chemical composition of the material (e.g. garnet) and the lattice parameters of layer and substrate. Implantation will change both the chemical composition and the lattice parameter. For example if one deposits a layer in which the magnetostriction were positive then the plane of magnetisation would be parallel to the film so no bubbles could develop. However implants change the situation and in selected regions we may rotate the plane of polarization to contain bubbles. (The converse is equally likely with different starting materials.) This type of change appears to be chemical in nature as implants with inert gases, to produce radiation damage, do not alter the plane of magnetisation (Wolfe et al., 1971).

(ii) To develop stable bubbles it is preferable to commence with a serrated domain pattern as this offers pinch off points which will provide the cylindrical bubbles (Bobeck, 1967). Once formed they may be maintained by a small steady field. Implantation via a mask can obviously form the necessary shaped domains. The boundary of the implant also forms a confining pattern for the bubble movement and Johnson et al. (1973) used implantation to define guide rails.

An interesting side effect noted by Johnson et al. (1973) is that the chemical stability of damaged garnets is greatly reduced. They implanted H, He or Ne ions and then etched the surface with phosphoric acid. Enhancements of up to 1000 times occurred at the implanted region so it was possible to etch conductor grooves in the garnet with the accuracy of the ion beam mask and ignore chemical widening of the groove. (More typical radiation enhanced etch rates found, for example of silica in hydrofluoric acid are only a factor of 5 greater than the normal etch rate.)

(iii) The formation of bubbles and the magnetic reversal of the domains is a function of the electron spin arrangement in the domain wall. After epitaxy there are a range of wall stabilities so that some domains invert more readily than others. This is undesirable and in particular one wishes to avoid "hard" boundaries. Rosencwaig (1972) suggested these result from a locked-in twisted arrangement of the electron spins which extend right through the layer from the substrate. Bobeck et al. (1972) avoided the problem by generating a multilayer structure. Since this is chemically, or epitaxially, difficult they found it was simpler to produce variations in the layer by ion implantation. An alternative approach of Wolfe and North (1972) was to suppress the hard bubbles by introducing a layer of radiation damage. This typically meant a hydrogen implant of 10^{16} ions cm^{-2} at an energy of 100keV, see also Hoshikawa and Harada (1974). The stability of the damaged garnets is different and hard bubbles were suppressed for subsequent heat treatments up to 300 to 1000°C in the various materials.

(iv) Bubble migration is normally controlled by current leads or magnetic layers superposed on the garnet layer. In ion implanted systems the bubble regions and the surrounding areas may be magnetised in orthogonal planes. This offers the possibility of control from the surrounding garnet layer without the addition of overlays.

(v) Finally the bubble velocity can be changed by implantation or radiation damage.

10.16 SIMULATION STUDIES OF DAMAGE IN REACTOR MATERIALS

Throughout the world immense sums of money are being invested in the construction of nuclear reactors. Alternative designs to optimise the extraction of energy from the fission process have produced a proliferation of reactor types with a wide range of coolant systems and constructional materials. The rate at which newer and higher powered designs are commissioned is too fast to allow prolonged life expectancy studies of earlier reactors to provide indications of all the problems that can result from large irradiation doses. In order to avoid an uncontrolled failure of a reactor or a premature shutdown it is essential to predict the long term radiation effects at the design stage. Such a statement is of course equally applicable to controlled fusion systems even though they are at a relatively early stage of development. The problem is difficult because there is defect production throughout the material. From the passage of neutron and fission fragments, or induced radioactivity, one continuously injects helium ions into the material in the form of alpha particles. Radiation damage levels will be high so that in the lifetime of a material in a reactor each atom may be displaced as often as 400 times. If failure occurs at lower levels of displacement per atom (dpa) then the reactor will be uneconomic. During the displacement event one forms both vacancies and interstitials and although some fraction of these recombine or move to sinks a significant number of the interstitials will precipitate into platelets or clusters and so leave a supersaturation of vacancies. If these also cluster then the "empty" spaces cause expansion of the material. The rate of formation of these empty volumes can be enhanced if there are nucleation sites such as gas atoms in the solid. Consequently the presence of helium (or oxygen, etc.) will develop a void structure and the materials will swell. Vacancy diffusion rates increase at high temperature so in high flux or high temperature reactors, such as breeder reactors, the result

10.16 SIMULATION STUDIES OF DAMAGE IN REACTOR MATERIALS

is exacerbated. Void formation could even be a consequence of a thermal annealing procedure to remove the stored energy in the lattice of the reactor materials. In a study of neutron irradiated type 316 stainless steel, a cladding material used in breeder reactors, Cawthorne and Fulton (1966) observed a void induced swelling in the material. Extrapolating this volume change over the reactor lifetime suggested a swelling of between 7 and 60% depending on the operating conditions of the reactor. This catastrophic prediction appeared to require a reactor lifetime to assess the actual volume change. Bullough *et al.* (1970) suggested that beyond an initial threshold dose the volume change caused by voids could be considered as a function of the rate

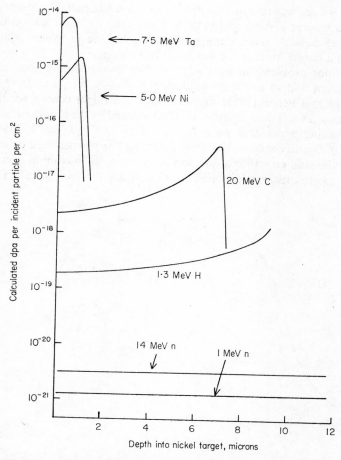

FIG. 10.14. The calculated displacements per atom produced by various particles in nickel.

of displacement (K), irradiation time (t), a parameter involving the sink concentration (S), and a term (F) dependent on the irradiation temperature and defect migration energies. They then wrote:

$$\frac{\Delta V}{\Delta} = Kt\,SF$$

Nelson and Mazey (1969) appreciated that the product term, Kt, could be achieved in a short time if the displacement rate could be increased above that available in a reactor. Ion implantation satisfied this requirement and Fig. 10.14 indicates that the passage of an energetic heavy ion can be some 10^6 more effective in producing displacements than the neutrons emitted during fission. It is therefore possible to realistically simulate the radiation damage conditions of a reactor lifetime within days rather than decades. Indeed the ability to survey possible reactor materials and solve the problems which ensue has been a major triumph for ion beam technology.

There are minor problems in simulation studies because some processes are rate dependent and so one must additionally modify the parameters K and F. Bullough and Perrin (1971) suggested that one might correct for the diffusion of vacancies and their equilibrium level induced by the bombardment flux by maintaining a constant ratio for the terms $K(T)/D_{vac}(T)$; where $D_{vac}(T)$ is the diffusion coefficient of the vacancies. The temperature dependence of F for the void growth is a function of flux and the maximum occurs at different temperatures, as indicated in Fig. 10.15 for two fluxes. The

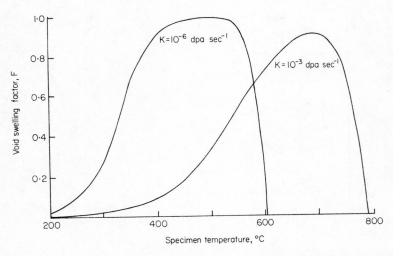

Fig. 10.15. The effect of particle flux on the temperature dependence of void swelling.

example is taken from the work of Brailsford and Bullough (1972) from which we see that to simulate reactor damage one must use implantation damage in samples held at a higher temperature than would exist in the reactor. Alternative thermal processes may add some uncertainty but for an empirical study the ion beam simulated damage is excellent.

One should note that in parallel studies of materials to be used in plasma walls of a fusion reactor there will be less of an extrapolation in flux effects than for fission reactors as the fusion processes is likely to operate as a series of high intensity pulse events.

Changes in the surface layers by blistering, surface flaking or failure of mechanical strength can all be simulated by ion implantation. But, as is apparent from Fig. 10.14 if the ions are massive then one may require accelerators in the MeV energy range. The choice of ion should probably be limited to the natural elements of the target and to achieve uniform and "bulk" effects the penetration should be a few microns. If this is not so then the surface will act as a sink for some damage products. Inevitably one will also consider if the implant is more mobile than the reactor induced defects and whether the implant is creating a new alloy. Additional subtleties in void formation may be resolved by simultaneous implants with a damaging

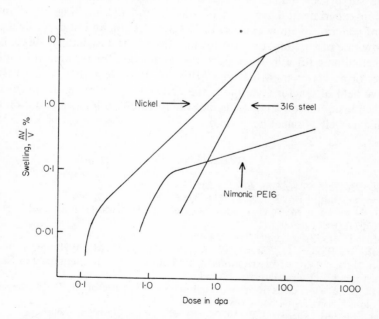

FIG. 10.16. The measured swelling produced by 20 MeV carbon atoms in Ni, 316SS, PE16 at 525°C. All samples contain 10 ppm He.

ion and helium injection. To satisfy these requirements implants have been made with energies up to 50 MeV with ions as heavy as nickel. A review of the subject was presented by Kulcinski (1974) and we shall now mention a few of the examples.

It was established that void formation can occur even in the absence of helium gas and void nucleation was observed in steels, Ni, Mo, V and Nb. One should note that other gas atoms may be effective in void nucleation and the presence of oxygen is common in many types of steel, particularly if titanium ions are included. However, if helium gas atoms are present there is a dramatic increase in the nucleation rate and density of voids. This is accompanied by an enhanced rate of swelling for a particular damage level.

An unexpected result was that several pure materials form voids in an ordered array, termed a superlattice, these have a structure, for example, b.c.c. or f.c.c., consistent with the crystal structure of the host lattice. At about the same damage level that the void superlattice appears in pure metals one finds a saturation in the swelling of the material. A saturation effect is also found in the swelling of type 316 stainless steel although in this case no superlattice occurs. The saturation is fortunate because the void induced swelling is slightly less of a problem than was originally feared.

In the surveys of materials suitable for reactor cladding the ion beam simulation experiments have been invaluable and the swelling noted in a variety of materials is shown in Fig. 10.16 (after Nelson and Mazey). Because of the low rate of swelling it is obvious that the nickel based alloy, PE16, is a suitable candidate for a high temperature reactor and for prolonged irradiations performs better than the type 316 steel. It is clear that in the vastly expensive field of reactor technology the ability to solve materials problems by ion beam simulation studies justifies the research on ion implantation on economic as well as scientific grounds.

REFERENCES

Agajanian, A. H. (1974). *Rad. Effects*, **23**, 73.
Anderson, D. B. and Boyd, J. T. (1971). *Appl. Phys. Lett.* **19**, 266.
Antill, J. E., Bennett, M. J., Dearnaley, G., Fern, F. H., Goode, P. D. and Turner, J. F. (1973). See Crowder (1973), 415.
Archard, J. F. (1953). *J. Appl. Phys.* **24**, 981.
Ashworth, V., Baxter, D., Grant, W. A., Proctor, R. P. M., Wellington, T. C. (1974). "Proceedings of International Conference on Ion Implantation in Semiconductors and Other Materials". Osaka.
Ashworth, V., Carter, G., Grant, W. A., Jones, P. D., Proctor, R. P. M., Sayegh, N. N. and Street, A. D. See Crowder (1973), 443.
Balkanski, M. and Lallemand, P. (1973). Eds. Photonics, Gauthier-Villars.
Bayly, A. R. (1973). *Rad. Effects*, **18**, 111.
Bayly, A. R. and Townsend, P. D. (1973). *J. Phys. D.* **6**, 1115.

REFERENCES

Bett, R. and Howlett, B. W. (1970). Cited by Dearnaley *et al.* (1973).
Bobeck, A. H. (1967). *Bell Syst. Tech. Journ.* **46**, 1901.
Bobeck, A. H., Blank, S. L. and Levinstein, H. J. (1972). *Bell Syst. Tech. Journ.* **51**, 1431.
Brailsford, A. D. and Bullough, R. (1972). *J. Nucl. Mat.* **44**, 121.
Brown, F. and Mackintosh, W. D. (1973). *J. Electrochem. Soc.* **120**, 1096.
Brown, W. L. (1971). "2nd Ion Implantation Conf. Garmisch," p. 430, Springer-Verlag.
Buckel, W. and Heim, G. (1974). See Picraux *et al.* (1974), p. 35.
Buckel, W. and Stritzker, B. (1974). See Picraux *et al.* (1974), p. 3.
Bullough, R., Eyre, B. L. and Perrin, R. C. (1970). *J. Nucl. Appl. and Tech.* **9**, 346.
Bullough, R. and Perrin, R. C. (1971). "Proc. Int. Conf. on Radiation Induced Voids in Metals", (Eds J. W. Corbett and L. C. Ianniello) (Conf-710601, 1972), p. 769.
Cawthorne, C. and Fulton, E. J. (1966). *Nature*, **216**, 575.
Chang, C. C. and Rose-Innes, A. C. (1970). "Proc. XII Int. Conf. on Low Temp. Phys." Kyoto.
Channin, D. J. (1971). *Appl. Phys. Lett.* **19**, 128.
Copeland, J. A. (1971). *J. Electron. Mater*, **1**, 420.
Cox, B. and Pemsler, J. P. (1968). *J. Nucl. Mater.* **28**, 73.
Crittenden, G. C., Townsend, P. D. and Townshend, S. E. (1974). *J. Phys. D.* **7**, 2397.
Crowder, B. L. (1973). (Ed.) "Ion Implantation in Semiconductors and Other Materials" Plenum, New York.
Crowder, B. L. and Tan, S. I. (1971). *IBM Tech. Disclosure Bull.* **14**.
Crozat, P., Adde, R., Chaumont, J., Bernas, H. and Zenatti, D. (1974). See Picraux *et al.* (1974), p. 27.
Dearnaley, G. (1974). See Picraux *et al.* (1974), p. 63.
Dearnaley, G., Freeman, J. H., Nelson, R. S. and Stephen, J. (1973). "Ion Implantation" North-Holland.
Dearnaley, G., Goode, P. D., Miller, W. S. and Turner, J. F. (1973). See Crowder (1973), p. 405.
Dearnaley, G., Goode, P. D. and Turner, J. F. (1971). Cited by Dearnaley *et al,* (1973).
Dearnaley, G., Hardacre, A. G. and Rogers, B. D. (1969). *Nucl. Inst. and Methods*, **71**, 86.
de Waard, H. (1972). "Mössbauer spectroscopy and its applications", IAEA, Vienna, p. 123.
de Waard, H. and Feldman, L. C. (1974). "Proc. of Conf. on Ion Implantation in Semiconductors and Other Materials", Osaka, p. 317.
Eer Nisse, E. P. (1971). *Appl. Phys. Lett.* **18**, 581.
Fiorito, G., Gasparrini, G. and Svelto, F. (1973). *Appl. Phys. Lett.* **23**, 448.
Fischer, K. (1970). *Springer Tracts Mod. Physics.* **54**, 1.
French, W. G. and Pearson, A. D. (1970). *Amer. Ceram. Soc. Bull.* **49**, 974.
Freyhardt, H. C., Taylor A. and Loomis, B. A. (1974). See Picraux *et al.* (1974), p. 47.
Fritzsche, C. R. (1973). *J. Electrochem. Soc.* **120**, 1603.
Garmire, E., Stoll, H., Yariv, A. and Hunsperger, R. G. (1972). *Appl. Phys. Lett.* **21**, 87.
Garvin, H. L., Garmire, E., Somekh, S., Stoll, H. and Yariv, A. (1973). *Appl. Optics*, **12**, 445.

Goell, J. E. (1973). *Appl. Optics*, **12**, 729.
Goell, J. E. and Standley, R. D. (1969). *Bell. Syst. Tech. J.*, **48**, 3445.
Goell, J. E., Standley, R. D., Gibson, W. M. and Rodgers, J. W. (1972). *Appl. Phys. Lett.* **21**, 72.
Grenness, M., Thompson, M. W., Cahn, R. W. (1974). *J. Appl. Electrochem.* **4**, 211.
Hartley, N. E. W. (1976). Proc. Int. Conf. on Applications of Ion Beams to Materials, I.O.P. To be published.
Hartley, N. E. W., Dearnaley, G., Turner, J. F. and Saunders, J. (1973). *AERE.* rept. R7580, see Picraux *et al.*, (1974), p. 123.
Hauffe, K. (1965). "Oxidation of Metals", Plenum Press.
Hayaski, N., Chang, H., Romankiv, L. T. and Krongelb, S. (1972). *IEEE Trans. Magn.* **MAG-8**, 16.
Hoshikawa, K. and Harada, H. (1974). "Proc. of Conf. on Ion Implantation in Semiconductors and Other Materials", Osaka.
Hughes, A. F. and Pells, G. P. (1974). *J. Phys. C. ibid* (1974). *Phys. Stat. Sol. A. IEEE Trans. Microwave Theory and Techniques* (1975), **23**, no. 1.
Johnson, W. A., North, J. C. and Wolfe, R. (1973). *J. Appl. Phys.* **44**, 4753.
Kaminow, I. P. and Carruthers, J. R. (1973). *Appl. Phys. Lett.* **22**, 326.
Kaminow, I. P., Carruthers, J. R., Turner, E. H. and Stulz, L. W. (1973). *Appl. Phys. Lett.* **22**, 540.
Kaminow, I. P., Ramaswamy, V., Schmidt, R. V. and Turner, E. H. (1974). *Appl. Phys. Lett.* **24**, 622.
Kaufmann, E. N., Raghavan, P., Raghavan, R. S., Krien, K., Ansaldo, E. J. and Naumann, R. A. (1974). "Proc. of Conf. on Ion Implantation in Semiconductors and Other Materials", Osaka, p. 379.
Kompfner, R. (1972). *Applied Optics.* **11**, 2412.
Kostka, A. and Kalbitzer, S. (1973). *Appl. Phys. Lett.* **23**, 705.
Kuhn, L., Dakss, M. L., Heidrich, P. F. and Scott, B. A. (1970). *Appl. Phys. Lett.* **17**, 265.
Kulcinski, G. L. (1974). See Picraux *et al.* (1974), p. 613.
Mackintosh, W. D. and Brown, F. (1974). See Picraux *et al.* (1974), p. 111.
Mandler, J. W., Moler, R. B., Raisen, E. and Rajan, K. S. (1973). *Thin Solid Films*, **19**, 165.
Marcatili, E. A. J. (1969). *Bell Syst. Tech. J.* **48**, 2103.
Marcuse, D. (1969). *Bell Syst. Tech. J.* **48**, 3233.
Marcuse, D. (Ed.) (1972). "Integrated Optics", IEEE press.
Marine, J. and Motte, C. (1973). *Appl. Phys. Lett.* **23**, 450.
Martin, W. E. and Hall, D. B. (1972). *Appl. Phys. Lett.* **21**, 325.
Mayer, J. W., Eriksson, L. and Davies, J. A. (1970). "Ion Implantation in Semiconductors", Academic Press.
Mayer, J. W. and Marsh, O. J. (1969). *Appl. Sol. State Science*, **1**, 239.
Meyer, O., Mann, H. and Phrilingos, E. (1974). See Picraux *et al.* (1974), p. 15.
Nelson, R. S. and Mazey, D. J. (1969). Proc. IAEA. Symp. on "Radiation Damage in Reactor Materials", Vienna, p. 157.
Nomura, K., Hirose, Y., Akasaka, Y., Horie, K. and Kawazu, S. (1974). "Int. Conf. on Ion Implantation in Semiconductor and Other Materials", Osaka, paper X-3, p. 172.
Picraux, S. T., Eer Nisse, E. P. and Vook, F. L. (Eds) (1974). "Applications of Ion Beams to Metals", Plenum.
Potter, E. C. (1956). "Electrochemistry", Cleaver-Hume.

REFERENCES

Presby, H. M. and Brown, W. L. (1974). *Appl. Phys. Lett.* **24**, 511.
Rand, M. J. and Standley, R. D. (1972). *Appl. Optics*, **11**, 2482.
Rosencwaig, A. (1972). *Bell. Syst. Tech. Journ.* **51**, 1440.
Saruwatari, M. and Izawa, T. (1974). *J. Appl. Phys. Lett.* **24**, 603.
Schineller, E. R., Flam, R. P. and Wilmot, D. W. (1968). *J. Opt. Soc. Amer.* **58**, 1171.
Schmidt, R. V., Kaminow, I. P. and Carruthers, J. R. (1973). *Appl. Phys. Lett.* **23**, 417.
Shubert, R. and Harris, J. H. (1971). *J. Opt. Soc. Amer.* **61**, 154.
Somekh, S., Garmire, E., Yariv, A., Garvin, H. L. and Hunsperger, R. G. (1973). *Appl. Phys. Lett.* **22**, 46.
Spalvins, T. (1971). *ASLE Trans.* **14**, 267.
Spitznagel, J. A., Fleischer, L. R. and Choyke, W. J. (1974). See Picraux *et al.* (1974), p. 87.
Standley, R. D., Gibson, W. M. and Rodgers, J. W. (1972). *Appl. Optics*, **11**, 1313.
Stroud, P. T. (1972). *Thin Solid Films*, **11**, 1.
Takagi, T., Yamada, I. and Kimura, H. (1974a). "Proc. of Conf. on Ion Implantation in Semiconductors and Other Materials", Osaka, p. 17.
Takagi, T., Yamada, I. and Sasaki (1974b). "Proc. of Conf. on Ion Implantation in Semiconductors and Other Materials", Osaka, p. 85.
Taylor, H. F., Martin, W. E., Hall, D. B. and Smiley, V. N. (1972). *Appl. Phys. Lett.* **21**, 95.
Tien, P. K. (1971). *Appl. Optics*, **10**, 2395.
Tien, P. K., Martin, R. J., Wolfe, R., Le Craw, R. C. and Blank, S. L. (1972a). *Appl. Phys. Lett.* **21**, 394.
Tien, P. K., Smolinsky, G. and Martin, R. J. (1972b). *Appl. Optics*, **11**, 637.
Tien, P. K., Ulrich, R. and Martin, R. J. (1970). *Appl. Phys. Lett.* **17**, 447.
Tove, P. A. and Chu, W. K. (1974). *The Catgut Acoustical Society Newsletter*, **22**, 5.
Trillat, J. J. and Hayman, P. (1961). "Le Bombardment Ionique", CNRS.
Uhlig, H. H. (1971). "Corrosion and Corrosion Control", John Wiley, New York
Ulrich, R. and Martin, R. J. (1971). *Appl. Optics*, **10**, 2077.
Wakelin, R. J. (1974). *Am. Rev. Mat. Sci.* **4**, 221.
Webb, A. P., Allen, L., Edgar, B. R., Houghton, A., Townsend, P. D. and Pitt, C. W. (1975). *J. Phys. D.,* **8**, 1567.
Wei, D. T. Y., Lee, W. W. and Bloom, L. R. (1973). *Appl. Phys. Lett.* **22**, 5.
Wei, D. T. Y., Lee, W. W. and Bloom, L. R. (1974). *Appl. Phys. Lett.* **25**, 329.
Wolfe, R. and North, J. C. (1972). *Bell Syst. Tech. Journ.* **51**, 1436.
Wolfe, R., North, J. C., Barns, R. L., Robinson, M. and Levenstein, H. J. (1971). *Appl. Phys. Lett.* **19**, 298.

Appendix I
RANGE–ENERGY DATA

The graphs presented in this appendix are intended to offer an estimate of the calculated projected range, R_P, and the standard deviation, ΔR_P, of R_P. We have chosen typical data from the tables of Johnson and Gibbons (1970) which are a computer evaluation of the LSS range–energy relations. Such calculations involve several assumptions and approximations in the choice of potential, contribution of electronic energy loss, and method of numerical integration. Comparison with experiment also involves such problems as radiation enhanced diffusion which modify the original implantation profile. Despite these limitations the calculated data offers a valuable first approximation to the range profile of implanted ions. The graphs include data for a range of Z_1 and Z_2 so that by interpolation and scaling for target density one can estimate ranges in other materials or ranges of different ions.

Not all authors use precisely the same computer programme so there are differences between the range estimates presented here (Johnson and Gibbons, (1970)) and those calculated by Wilson and Brewer (1973), or Smith (1971). Smith's data is reproduced in tabular form as an appendix to the book by Dearnaley et al. (1973).

The density of the materials considered in the graphs is:—

Aluminium	2·70 gm cm^{-3}
Cadmium Telluride	6·20
Diamond	3·51
Germanium	5·35
Gold	19·3
Silica	2·27

REFERENCES

Dearnaley, G., Freeman, J. H., Nelson, R. S. and Stephen, J. (1973). "Ion Implantation", North Holland.

Johnson, W. S., and Gibbons J. F. (1970). "Projected Range Statistics in Semiconductors", distributed by Stanford University Bookstore (1969); *ibid* second edition distributed by Halstead Press (Wiley).

Smith, B. J. (1971). A.E.R.E. report R 6660.

Wilson, R. G. and Brewer, G. R. (1973). "Ion Beams", Wiley.

APPENDIX I

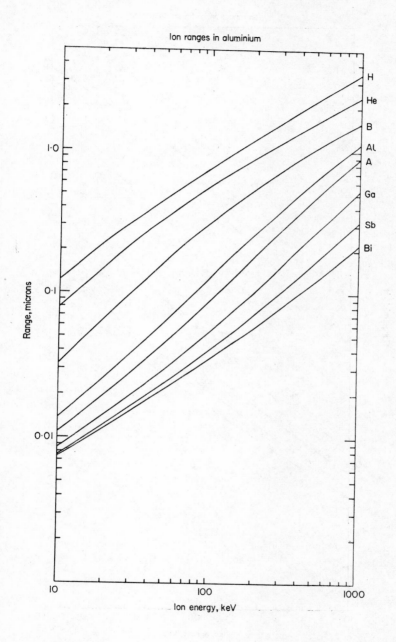

APPENDIX I

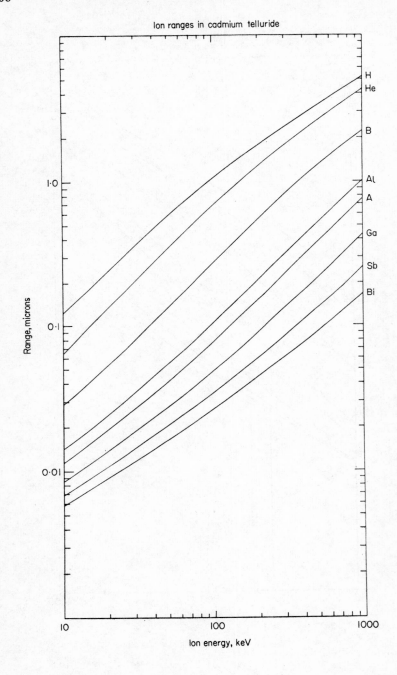

Ion ranges in cadmium telluride

APPENDIX I

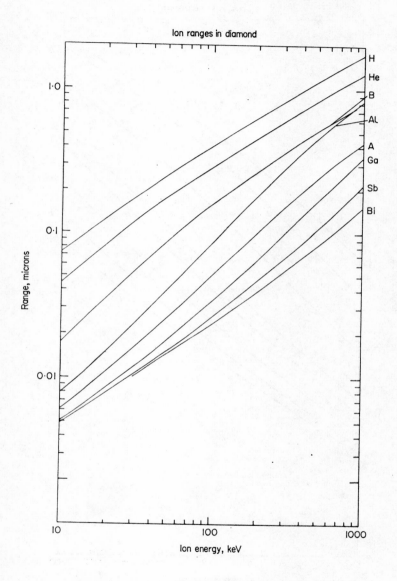

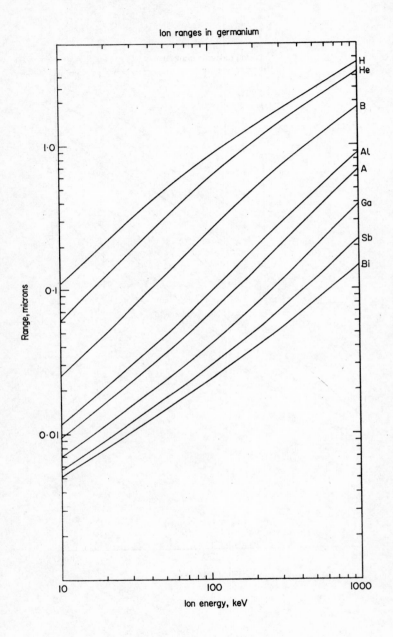

APPENDIX I

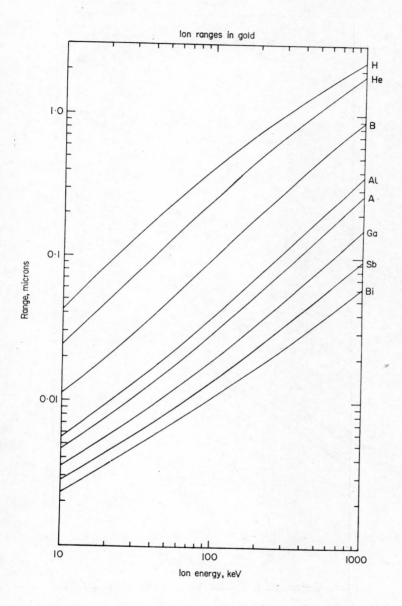

APPENDIX I

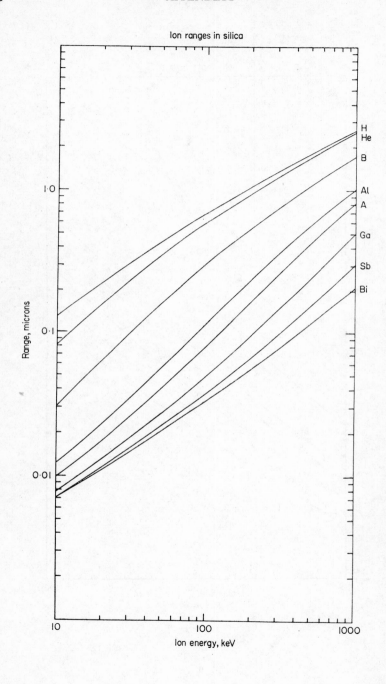

Ion ranges in silica

APPENDIX I

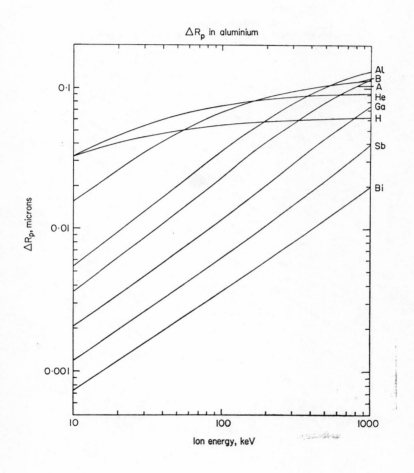

ΔR_p in aluminium

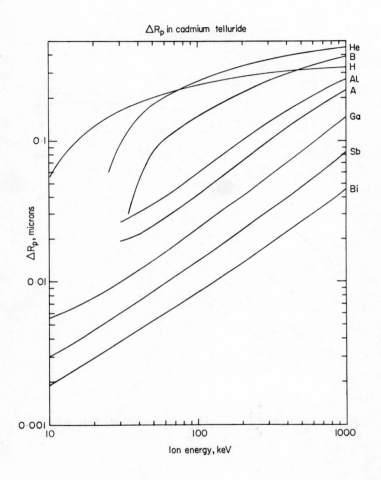

APPENDIX I

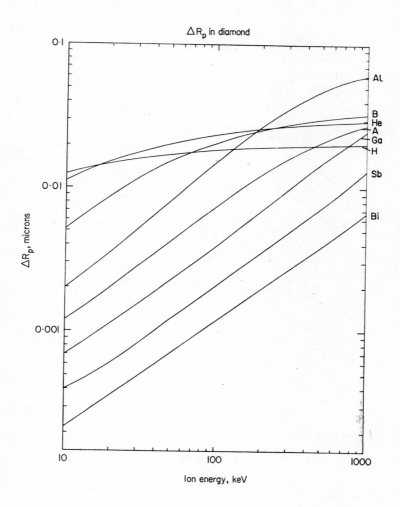

APPENDIX I

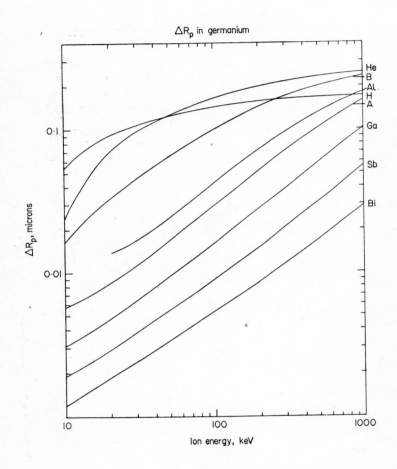

APPENDIX I

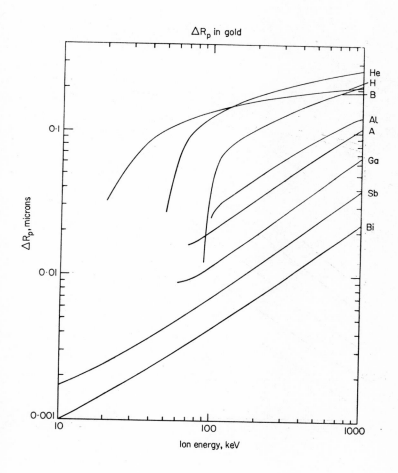

ΔR_p in gold

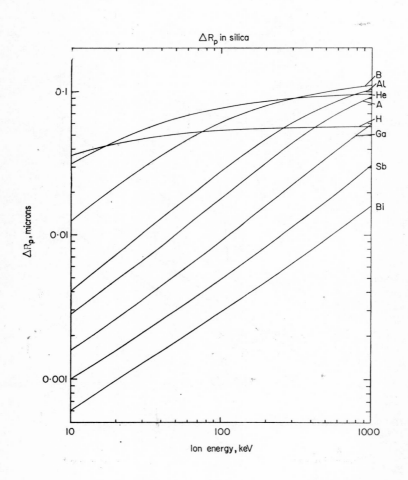

Appendix II

TABLE OF Z, A AND NATURAL ABUNDANCE OF THE ISOTOPES

Element	Z	Mass	Natural abundance (%)
H	1	1·008	99·985
		2·014	0·015
He	2	3·016	$<10^{-3}$
		4·003	~ 100
Li	3	6·015	7·42
		7·016	92·58
Be	4	9·012	100
B	5	10·013	19·6
		11·009	80·4
C	6	12·000	98·89
		13·003	1·11
N	7	14·003	99·63
		15·000	0·37
O	8	15·995	99·759
		16·999	0·037
		17·999	0·204
F	9	18·998	100
Ne	10	19·992	90·92
		20·994	0·257
		21·991	8·82
Na	11	22·990	100

APPENDIX II

Element	Z	Mass	Natural abundance (%)
Mg	12	23·985	78·70
		24·986	10·13
		25·983	11·17
Al	13	26·982	100
Si	14	27·977	92·21
		28·976	4·70
		29·974	3·09
P	15	30·974	100
S	16	31·972	95·0
		32·971	0·76
		33·968	4·22
		35·967	0·014
Cl	17	34·969	75·53
		36·966	24·47
Ar	18	35·968	0·337
		37·963	0·063
		39·962	99·60
K	19	38·964	93·10
		~40	0·0118
		40·962	6·88
Ca	20	39·963	96·97
		41·959	0·64
		42·959	0·145
		43·955	2·06
		45·954	0·0033
		47·952	0·18
Sc	21	44·956	100
Ti	22	45·953	7·93
		46·952	7·28
		47·948	73·94
		48·948	5·51
		49·945	5·34
V	23	49·947	0·24
		50·944	99·76

APPENDIX II

Element	Z	Mass	Natural abundance (%)
Cr	24	49·946	4·31
		51·941	83·76
		52·941	9·55
		53·939	2·38
Mn	25	54·938	100
Fe	26	53·940	5·82
		55·935	91·66
		56·935	2·19
		57·933	0·33
Co	27	58·933	100
Ni	28	57·935	67·88
		59·933	26·23
		60·931	1·19
		61·928	3·66
		63·928	1·08
Cu	29	62·930	69·09
		64·928	30·91
Zn	30	63·929	48·89
		65·926	27·81
		66·927	4·11
		67·925	18·57
		69·925	0·62
Ga	31	68·926	60·4
		70·925	39·6
Ge	32	69·924	20·52
		71·922	27·43
		72·923	7·76
		72·921	36·54
		75·921	7·76
As	33	74·922	100
Se	34	73·923	0·87
		75·919	9·02
		76·920	7·58
		77·917	23·52
		79·917	49·82
		81·917	9·19

APPENDIX II

Element	Z	Mass	Natural abundance (%)
Br	35	78·918	50·54
		80·916	49·46
Kr	36	77·920	0·35
		79·916	2·27
		81·914	11·56
		82·914	11·55
		83·912	56·90
		85·911	17·37
Rb	37	84·912	72·15
		∼87	27·85
Sr	38	83·913	0·56
		85·909	9·86
		86·909	70·2
		87·906	82·56
Y	39	88·905	100
Zr	40	89·904	51·46
		90·905	11·23
		91·905	17·11
		93·906	17·40
		95·908	2·80
Nb	41	92·906	100
Mo	42	91·906	15·84
		93·905	9·04
		95·905	16·53
		96·906	9·46
		97·906	23·78
		99·908	9·13
Tc	43		
Ru	44	95·908	5·51
		97·906	1·87
		98·906	12·72
		99·903	12·62
		100·904	17·07
		101·904	31·61
		103·906	18·58
Rh	45	102·905	100

APPENDIX II

Element	Z	Mass	Natural abundance (%)
Pd	46	101·905	0·96
		103·904	10·97
		105·905	22·23
		105·903	27·33
		107·903	26·71
		109·905	11·81
Ag	47	106·904	51·82
		108·905	48·18
Cd	48	105·907	1·22
		107·905	0·88
		109·903	12·39
		110·904	12·75
		111·903	24·07
		112·905	12·26
		113·904	28·86
		115·905	7·58
In	49	112·904	4·28
		114·904	95·72
Sn	50	111·904	0·96
		113·903	0·66
		114·904	0·35
		115·902	14·30
		116·903	7·61
		117·902	24·03
		118·903	8·58
		119·902	32·85
		121·903	4·92
		123·905	5·94
Sb	51	120·904	57·25
		122·904	42·75
Te	52	119·905	0·089
		121·903	2·46
		122·904	0·87
		123·903	4·61
		124·904	6·99
		125·903	18·71
		127·905	31·79
		129·907	34·48
I	53	126·904	100

APPENDIX II

Element	Z	Mass	Natural abundance (%)
Xe	54	123·906	0·096
		125·904	0·090
		127·903	1·92
		128·905	26·44
		129·904	4·08
		130·905	21·18
		131·904	26·89
		133·905	10·44
		135·907	8·87
Cs	55	132·904	100
Ba	56	129·906	0·101
		131'906	0·097
		133·904	2·42
		134·906	6·59
		135·904	7·81
		136·906	11·32
		137·905	71·66
La	57	137·907	0·089
		138·906	99·9111
Ce	58	135·907	0·193
		137·906	0·250
		139·905	88·48
		141·909	11·07
Pr	59	140·907	100
Nd	60	141·908	27·11
		142·910	12·17
		143·910	23·85
		144·912	8·30
		145·913	17·22
		147·917	5·73
		149·921	5·62
Pm	61		
Sm	62	143·912	3·09
		146·915	14·97
		147·915	11·24
		148·917	13·83
		149·917	7·44
		151·920	76·72
		153·922	22·71

APPENDIX II

Element	Z	Mass	Natural abundance (%)
Eu	63	150·920	47·82
		152·921	52·18
Gd	64	151·920	0·200
		153·921	2·15
		154·923	14·73
		155·922	20·47
		156·934	15·68
		157·924	24·87
		159·907	21·90
Tb	65	158·925	100
Dy	66	155·924	0·052
		157·924	0·090
		159·825	2·29
		160·927	18·88
		162·927	25·53
		162·928	24·97
		163·929	28·18
Ho	67	164·930	100
Er	68	161·929	0·136
		163·929	1·56
		165·930	33·41
		166·932	22·94
		167·932	27·07
		169·936	14·88
Tm	69	168·934	100
Yb	70	167·934	0·135
		169·935	3·03
		170·937	14·31
		171·937	21·82
		172·938	16·13
		173·939	31·84
		175·943	12·73
Lu	71	174·941	97·41
		~176	2·59

APPENDIX II

Element	Z	Mass	Natural abundance (%)
Hf	72	173·940	0·18
		175·944	5·20
		176·944	18·50
		177·944	27·14
		178·946	13·75
		179·947	35·24
Ta	73	179·948	0·0123
		180·948	99·988
W	74	179·947	0·14
		181·948	26·41
		182·950	14·40
		183·951	30·64
		185·954	28·41
Re	75	184·953	37·07
		186·956	62·93
Os	76	183·953	0·018
		185·954	1·59
		186·956	1·64
		187·956	13·3
		188·959	16·1
		189·959	26·4
		191·961	41·0
Ir	77	190·961	37·3
		192·963	62·7
Pt	78	189·960	0·0127
		191·961	0·78
		193·963	32·9
		194·965	33·8
		195·965	25·3
		197·968	7·21
Au	79	196·967	100
Hg	80	195·966	0·146
		197·967	10·02
		198·968	16·84
		199·968	23·13
		200·970	13·22
		201·971	29·80
		203·974	6·85

Element	Z	Mass	Natural abundance (%)
Tl	81	202·972	29·50
		204·975	70·50
Pb	82	203·973	1·48
		205·975	23·6
		206·976	22·6
		207·977	52·3
Bi	83	208·980	100
Po	84		
At	85		
Rn	86		
Fr	87		
Ra	88		
Ac	89		
Th	90		
Pa	91		
U	92	234·041	0·0057
		235·044	0·72
		238·051	99·27

Appendix III

ELECTRONICS FOR SURFACE ANALYSIS

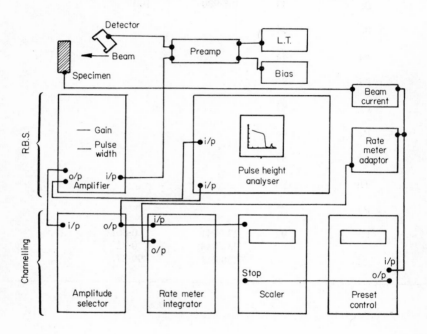

1. Block diagram of the electronics for Rutherford backscattering and channelling.

APPENDIX III

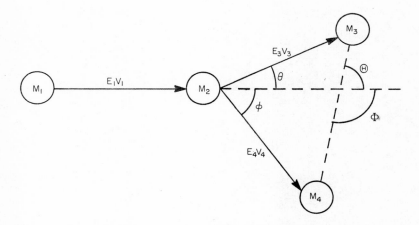

2. Nuclear reaction kinematics (see equation 8.20).

Index

A

abrasion resistance, 290
accelerators, 148 *et seq.*
adhesion, 34, 93, 270, 288
advantages of ion implantation, 3, 247, 263, 281
alkali halides, 72, 76, 77, 206, 235, 236, 264
alloys, 84, 91, 100, 114, 204 *et seq.*, 266, 273, 292
amorphous solids, 25, 38, 65, 115
analysis of ion beams, 166 *et seq.*
analysis with ion beams, 181 *et seq.*
angular dependence of sputtering yield, 126
angular momentum, 16
annealing, 33, 69, 71, 88 *et seq.*
anodic stripping, 37, 89, 272
anodisation, 271
aspheric lenses, 248
Auger transitions, 36, 218, 225
axial channelling, 51, 56, 60

B

backscattering, *see* Rutherford scattering, ion scattering
beam leads, 253
beam scanning, *see* ion beam transport
biological applications, 235, 258
blistering, 85, 299
blocking, 46, 201
blood cells, 258
bubbles, 84, 137, 217
Burgers vector, 81, 82

C

CAMECA, 230
capillary, machining of, 249
catalysis, 266, 267, 270
channelling, 35, 38, 45 *et seq.*, 105, 127, 198 *et seq.*, 277
channelling yields, 49
channel potentials, 51
characteristic energy, 27
charge exchange, 7, 24
chemical effects, 3, 259, 266, 289
chemical etching, 37, 79, 132, 252, 257
chemical dissolution, 132
coatings, *see* sputter deposition
collision cascade, 40, 65, 112, 126, 137
colour centres, 72, 76
compounds, ion ranges in, 31, 192
computer simulations, 12, 40
concentration profiles, 31
cone formation, 129, 135, 143
continuum potential, 53
corrosion, 194, 213, 217, 269 *et seq.*
cosputtering, 144
Coulomb barrier, 211
couplers, 257, 280
critical angle, 49, 58, 59, 201
cross section, 16, 22, 186, 219, 224
crowdion, 67, 68, 77

D

damage profiles, 26
dechannelling, 48, 62, 201
depth resolution, 98, 103

desorption, 228
detection sensitivity 203, 216, 222, 239
diffraction gratings, 257, 284
diffusion, 88 *et seq.*, 105, 289, 296
diffusion coefficient, 32
diode, 255
dislocations, 79 *et seq.*, 142
dislocation pinning, 83
displacement energy, 70, 74, 77
divacancy, 69
domains, 295
dose measurement, 176 *et seq.*
double alignment, 201
duoplasmatron source, 153, 164

E

edge dislocation, 80
elastic collisions, 7
electrochemistry, 266, 267
electronegativity, 205, 270
electron excitation, 7
electron microscopy, samples for, 251
electron nuclear double resonance (ENDOR), 71, 275
electron shell effects, 13, 21, 62, 206, 224
electron spin resonance (ESR), 71, 275
electron suppression, 177
electronic effects in defect production, 75
electronic energy loss, 21, 62
encapsulants, 231, 263
energy dispersive X-ray analysis (EDAX), 214
energy loss, 7 *et seq.*
enhanced diffusion, 89
embrittlement, hydrogen, 217
erosion of surface, 134

F

Fick's law, 88
field effect transistors, 256
figuring, 248
fission fragments, 79

flux peaking, 57, 202, 277
formation energy, 66 *et seq.*
Frenkel defect, 68
friction, 197, 287 *et seq.*
fusion reactor, 217, 296

G

gallium arsenide, 22
garnets, 5, 295
geological specimens, 182, 235, 251
gettering, 90, 106
glancing incidence analysis, 98, 198
glass, machining of, 248, 250
grain boundary diffusion, 93, 95
goniometer, 173, 174
guides, 281

H

hardness changes, 290
hollow cathode source, 161
hydrogen detection, 217
hyperfine interactions, 203, 275

I

impulse approximation, 17, 50
inelastic collisions, 7 *et seq.*,
inner shell effects, 39
insulators, 71, 89, 95, 228, 232, 235
integrated circuits, 93, 253, 263
integrated optics, 256 *et seq.*, 280 *et seq.*, 287
interatomic potentials, 11, 12, 52, 63, 120, 198
interdiffusion, 94
interstitial, 66 *et seq.*, 105, 202, 277
interstitial diffusion, 105
ion beam etching, *see* sputter etching
ion beam polishing, 37, 132, 249 *et seq.*
ion beam transport, 148 *et seq.*, 166, 171, 175

INDEX 331

ion induced X-rays, 218
ionisation efficiency, 157
ionisation potential, 157
ion microprobe, 230
ion neutralisation spectroscopy (INS), 238, 239
ion plating, 145, 288
ion ranges, 7 et seq.
ion scattering, 181 et seq.
ion scattering spectroscopy (ISS), 238
ion sources, 149 et seq.
isotope separation, 161 et seq., 170
isotopic abundance, 317 et seq.

K

knock-on events, 231

L

laser, 284
laser beam damage, 250
laser feedback, 258
lateral range straggling, 31
lattice location, 202, 271, 275
lithium niobate, 214, 253, 256, 282, 285
liquid residues, 235
low energy scattering, 198
L.S.S. range theory, 7 et seq.
lubricating oxides, 288, 292

M

machining by sputtering, 111, 248, 257
magnetic analyser, 166, 167
magnetic bubbles, 253, 294 et seq.
magnetic suppression, 177
masks, 34, 248, 253, 257, 259
microanalysis, 213
microbeam, 98, 196, 215
microwave devices, 34, 255
migration energy, 69
minimum yield, 49, 63, 201

momentum approximation, 17, 50
momentum conservation, 8, 211, 327
Monte Carlo calculations, 40
moon, 182, 251
Mossbauer experiments, 203, 276
multiply charged ions, 164

N

negative ion sources, 150
neutral beams, 166, 233
nitrides, 206, 266, 269
nuclear energy level diagrams, 212
nuclear reactions, 47, 56, 197, 210, 277, 278

O

optical emission, 233, 235
oxidation, 206, 269 et seq.
oxides as semiconductors, 271
oxygen detection, 194, 207

P

paraxial approximation, 175
passivation, 266, 269, 271, 273
pattern resolution, 255
phosphors, 263 et seq.
photodiodes, 106, 266
photoresists, 34, 254, 257
piezoelectric tranducers, 253
pits, 136, 141
planar channelling, 52
plasma arc source, 150
polarization curves, 273
polariser, infrared, 256
polishing by ion beams, 111, 248 et seq.
pollution, 187, 220
polymers, 258
potentiostatic measurements, 272
powder samples, 194
power law potentials, 18, 21, 120

precipitation, 84, 103
projected range, 25, 304 *et seq.*
proton induced X-rays, 219, 221

Q

quadrupole analyser, 166
quartz resonators, 252
quasi-channelling, 54

R

radiation damage, 26, 65 *et seq.*, 142, 295, 296
radiation dosimetry, 265
radiation enhanced diffusion, 34
radioactive implants, 37, 46, 59, 105
radiofrequency sources, 151
range estimates, 7 *et seq.*, 304 *et seq.*
range straggling, 7 *et seq.*, 108, 206, 304 *et seq.*
reactor materials, 296
reduced energy, 22
reduced range, 27
refractive index, 282
replacement collision sequence, 76, 112
resonant reactions, 207 *et seq.*
resolution, Rutherford backscattering, 98, 103, 192
resolution, isotope separation, 172
Rutherford scattering, 19, 90, 98, 107, 181 *et seq.*, 186, 271, 277, 279, 326

S

sapphire rods, polishing of, 250
scattering events, 15
Schottky defect, 68
screw disclocation, 80, 81
secondary electrons, 176
secondary ion mass spectroscopy (SIMS), 225, 232
semiconductors, 2, 262 *et seq.*
simulation of reactor radiation, 296, 299

solid lubricants, 289
space charge, 175
sputter deposition, 143, 258, 259, 282, 288
sputter etching, 227, 231, 252, 253, 256, 295
sputtering, 41, 77, 111 *et seq.*, 227, 247 *et seq.*
sputtering, alkali halides, 143
sputtering, insulators, 144
sputtering yield, 114, 117, 121 *et seq.*, 178, 232
steel, 205, 216, 222, 230, 270, 290, 300
stoichiometry, 96, 267
stopping power, 7 *et seq.*, 99, 189
straggling, *see* range straggling
strain, 4, 249, 288
stress, 95, 96, 206, 267, 269, 292
string potential, 51
substitutional atoms, 202, 204, 277
superconductivity, 258, 292 *et seq.*
surface acoustic waves, 256, 286
surface barrier detector, 183
surface binding, 116, 121, 125
surface compositional analysis by neutral and ion impact radiation (SCANIIR), 225, 231, 233
surface contamination, 216, 225
surface ionisation source, 157
surface preparation, 58, 128, 144
surface roughness, 101, 128, 195
surface scattering, 238 *et seq.*
swelling, 85, 296 *et seq.*

T

thermal spike, 65, 127
thin films, 93, 94, 143, 185, 266, 287
thinning, 251, 253
tooth enamel, 278
topography, 128 *et seq.*, 136
transmission experiments, 48
transverse energy, 54
two body collisions, 115

INDEX

U

uniformity of profile, 32
uranium dioxide, 206

V

vacancies, 66 *et seq.*, 203, 296 *et seq.*
valence band analysis, 238
varnish layers, 279
velocity filter, 170
velocity spectrum of sputtered atoms, 126, 144
vibratory polishing, 37
violins, 279
voids, 79, 84, 293, 296 *et seq.*
void lattice, 300

W

waveguides, optical, 257 *et seq.*, 281, 284, 287

wear, 287 *et seq.*, 290
work function, 157, 177, 227, 229
work hardening, 84

X

x-ray cross section, 219, 221
x-ray emission, 36, 47, 218, 224
x-ray photoelectron spectroscopy, 227, 229

Y

yttrium 270, 271

Z

zircaloy, 216, 269
zirconium, 208

192037

RETURN TO ➡ **PHYSICS LIBRARY**
351 LeConte Hall   642-3122

LOAN PERIOD 1 1-MONTH	2	3
4	5	6

ALL BOOKS MAY BE RECALLED AFTER 7 DAYS
Overdue books are subject to replacement bills

DUE AS STAMPED BELOW

APR 1 4 1996		
MAR 1 5 1996 Rec'd UCB PHYS	SEP 1 5 2011	
JUL 1 2 1996		
AUG 2 9 1996 Rec'd UCB PHYS		
APR 3 0 1999 DEC 2 0 1999		
NOV 2 7 2005 MAR 1 7 2007		
APR 2 8 2010		
MAY 0 2 2011		

FORM NO. DD 25

UNIVERSITY OF CALIFORNIA, BERKELEY
BERKELEY, CA 94720